BASIC PURE
MATHEMATICS II

VNR NEW MATHEMATICS LIBRARY

under the general editorship of

J. V. ARMITAGE
Professor of Mathematics
University of Nottingham

N. CURLE
Professor of Applied Mathematics
University of St. Andrews

The aim of this series is to provide a
reliable modern coverage of those mainstream
topics that form the core of mathematical
instruction in universities and comparable
institutions. Each book deals concisely with
a well-defined key area in pure or applied
mathematics or statistics. Many of the
volumes are intended not solely for students
of mathematics, but also for engineering and
science students whose training demands a
firm grounding in mathematical methods.

Basic Pure
Mathematics II

by

A. S.-T. LUE, B.Sc., Ph.D.

Lecturer in Mathematics
University of London, King's College

VAN NOSTRAND REINHOLD COMPANY

New York - Cincinnati - Toronto - London - Melbourne

© A. S.-T. Lue 1974

ISBN 0 442 30048 4 cloth
ISBN 0 442 30049 2 paper

Library of Congress Catalogue Card
Number: 74-3595

Published by Van Nostrand Reinhold
Company Limited, 25-28 Buckingham Gate,
London SW1E 6LQ

Printed in Great Britain by William Clowes
& Sons Limited London, Colchester and
Beccles

Preface

This book is intended as a basic text for first year science undergraduates. It is based on a course given at King's College, London over the past five years to single honours science and joint honours mathematics students. That course was covered in thirty-six lectures during the second half of the academic year, and was preceded by Basic Pure Mathematics I (see the book of the same name in this series by Professor J. V. Armitage).

The division of material between Basic Pure I and Basic Pure II was roughly as follows.

I. Vectors in three dimensions, scalar and vector product, triple product with application to geometry of lines and planes.

Differentiation of vectors, brief mention of vector fields. Functions of several variables (as scalar-valued functions of vectors). Basic linear algebra in R^3. Matrices via linear transformations, determinants via transforming the triple product. Applications to linear equations (mainly 3×3).

Partial differentiation, differentiability, gradient, level curves. Change of variables, chain rule.

Revision of integration (integrals based on area). Multiple integrals and evaluation by repeated integration.

II. Complex numbers, de Moivre's theorem and applications.

Basic sequences and series and simple convergence tests. Exponential and logarithmic functions. Limits and continuous functions.

Mean value theorems in one and more variables. Taylor's theorem with remainder. Maxima and minima (including Lagrange multipliers). Expansion in series.

Differential equations: ordinary linear with constant coefficients, linear homogeneous, etc. Selected partial differential equations with special techniques.

Fourier series.

Not unnaturally, there was a certain amount of overlap of material in the two courses. This we found to be beneficial to the students, not only as revision, but also to emphasize certain aspects of the course.

We have endeavoured to maintain a reasonable degree of mathematical rigour throughout the course. Specialist mathematicians will therefore find the treatment of their favourite topics to be quite acceptable. The objective has been to make the exposition as clear and rigorous as possible, and then to illustrate the work by carefully chosen worked examples and exercises.

My particular thanks go to my colleague Dr. J. A. Tyrrell, to whom I am indebted for many constructive criticisms and comments.

A. S.-T. Lue

Contents

CHAPTER 1

Complex Numbers

1.1 Definitions

A *complex number* is an ordered pair (x, y) of real numbers x and y. The ordering is important here, and so the complex number $(3, 4)$ is different from the complex number $(4, 3)$. In fact, (x, y) and (x', y') are equal only if $x = x'$ and $y = y'$.

If we denote the complex number (x, y) by z, then we say that "*x is the real part of z*", and write $x = \text{Re}(z)$ and "*y is the imaginary part of z*", and write $y = \text{Im}(z)$.

The symbol $\mathbb{C}$ is used to denote the set of complex numbers, and in $\mathbb{C}$ *addition* and *multiplication* are defined as follows:

$$(x, y) + (x', y') = (x + x', y + y'),$$
$$(x, y)(x', y') = (xx' - yy', xy' + x'y).$$

EXERCISE (i)
Prove the following identities.

$$z + z' = z' + z, \qquad zz' = z'z \qquad \text{(Commutative laws)};$$
$$z + (z' + z'') = (z + z') + z'',$$
$$z(z'z'') = (zz')z'' \qquad \text{(Associative laws)};$$
$$z(z' + z'') = zz' + zz'' \qquad \text{(Distributive law)}.$$

The complex number $(0, 0)$ behaves, with respect to addition, like the *zero* in the real numbers, while $(1, 0)$ behaves, with respect to multiplication, like the *unit* in the real numbers. That is to say, for every complex number (x, y) we have

$$(x, y) + (0, 0) = (x, y),$$
$$(x, y)(1, 0) = (x, y).$$

If $z = (x, y)$, then quite clearly

$$(x, y) + (-x, -y) = (0, 0),$$

and so $(-x, -y)$ is the *additive inverse* of z, which we therefore denote by $-z$. Furthermore, if $z \neq (0, 0)$ then $x^2 + y^2 \neq 0$, and

$$(x, y)\left(\frac{x}{x^2 + y^2}, \frac{-y}{x^2 + y^2}\right) = (1, 0).$$

1

Thus $(x/(x^2 + y^2), -y/(x^2 + y^2))$ acts as the *multiplicative inverse* of z, and we denote it by z^{-1} or $1/z$.

In algebra, systems which have properties such as those we have described for the complex numbers are called *fields*, and so we often speak of the *field of complex numbers* (as indeed we speak of the *field of real numbers*).

EXERCISE (ii)
Find the (multiplicative) inverses of the complex numbers

(a) $(0, 1)$,   (b) $(1, 1)$,   (c) $\left(-\dfrac{1}{2}, \dfrac{\sqrt{3}}{2}\right)$,   (d) $(3, 4)$.

Quite clearly, the complex numbers of the form $(x, 0)$ are in one-one correspondence with the real numbers x. Since also

$$(x, 0) + (x', 0) = (x + x', 0),$$
$$(x, 0)(x', 0) = (xx', 0)$$

then we may identify the complex number $(x, 0)$ with the real number x, and regard the real numbers as a subset of $\mathbb{C}$. With this identification we see that, if a, b, c are real numbers then

$$a(b, c) = (a, 0)(b, c) = (ab, ac).$$

Therefore

$$(x, y) = (x, 0) + (0, y) = x + y(0, 1).$$

We denote the complex number $(0, 1)$ by the symbol i, and note that

$$i^2 = (0, 1)(0, 1) = (-1, 0) = -(1, 0) = -1.$$

In other words, *every complex number (x, y) can be written in the form $x + iy$, where $i^2 = -1$.*

The *conjugate* of z, where $z = x + iy$, is defined to be $x - iy$, and is denoted by $\bar{z}$. It follows from the definition that

$$z + \bar{z} = 2x, \qquad z - \bar{z} = 2iy, \qquad z\bar{z} = x^2 + y^2,$$

and so

$$\operatorname{Re}(z) = \tfrac{1}{2}(z + \bar{z}), \qquad \operatorname{Im}(z) = \frac{1}{2i}(z - \bar{z}) = -\frac{i}{2}(z - \bar{z}).$$

EXERCISE (iii)
Prove the following relations

(a) $\overline{z + z'} = \bar{z} + \bar{z'}$,   (b) $\overline{zz'} = \bar{z}\,\bar{z'}$,   (c) $\overline{z/z'} = \bar{z}/\bar{z'}$.

It is very often convenient to express the inverse of a complex number z in the form $z^{-1} = \bar{z}/(z\bar{z})$, and this enables us to compute the inverse a bit more readily. Thus, for example,

$$(1 + 2i)^{-1} = \frac{1 - 2i}{(1 + 2i)(1 - 2i)} = \frac{1 - 2i}{1 + 4} = \frac{1}{5} - \frac{2}{5}i.$$

Similarly,

$$\frac{1 + 7i}{3 + i} = \frac{(1 + 7i)(3 - i)}{(3 + i)(3 - i)} = \frac{10 + 20i}{9 + 1} = 1 + 2i.$$

EXERCISE (iv)
Express the following complex numbers in the form $x + iy$

(a) $\dfrac{-1 + 5i}{2 + 3i}$,   (b) $\dfrac{7 + 7i}{i}$,   (c) $\dfrac{2 + 5i}{5 - 2i}$,   (d) $\dfrac{34 - 2i}{3 - 7i}$.

The *absolute value* or *modulus* of a complex number $x + iy$ is defined to be $\sqrt{(x^2 + y^2)}$, and denoted by $|x + iy|$. Note that this is always a non-negative real number, and is equal to 0 only when the complex number is itself 0. Clearly

$$|z|^2 = |-z|^2 = |\bar{z}|^2 = x^2 + y^2.$$

EXERCISE (v)
Prove that the following relations hold

(a) $|zz'| = |z||z'|$,   (b) if $z' \neq 0$, then $\left|\dfrac{z}{z'}\right| = \dfrac{|z|}{|z'|}$.

EXERCISE (vi)
Evaluate

(a) $|2 + i|$,   (b) $|4 + 3i|$,   (c) $|(2 + i)(4 + 3i)|$,

(d) $\left|\dfrac{4 + 3i}{2 + i}\right|$.

1.2   The Argand diagram

Once we have chosen two mutually perpendicular axes and real scales on these axes, then every point in the plane can be represented by an ordered pair (x, y) of real numbers, the rectangular (or cartesian) coordinates of the point. Conversely, every such ordered pair (x, y), and hence every complex number, determines a unique point in the plane. When we are considering the representation of complex numbers by points in the plane, then the plane is called the *complex plane* or *Argand diagram*. The point in the Argand diagram which represents the complex number z is often referred to as "the point z".

Real numbers (i.e., complex numbers of the form $(x, 0)$) are represented by points on the x-axis, and so this is also called the *real-axis*. *Purely imaginary numbers* (i.e., complex numbers of the form $(0, y)$) are represented by points on the y-axis, and this is called the *imaginary-axis*.

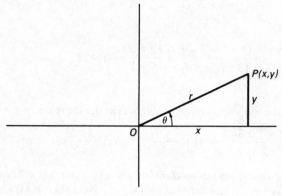

Figure 1

If P is the point z, where $z = x + iy$, then we see from Fig. 1 that

$$x = r \cos \theta, \qquad y = r \sin \theta$$

where $r = \sqrt{(x^2 + y^2)} = |z|$. Thus the modulus of z is the distance of the point z from the origin. The angle θ which OP makes with the positive real-axis is called the *argument* or *amplitude* of z, and is denoted by *arg z* (or *am z*). Clearly, it has a meaning only if $z \neq 0$. To avoid ambiguity, we shall adopt the convention that $0 \leqslant \theta < 2\pi$. The complex number z can therefore be written in the form

$$z = x + iy = r (\cos \theta + i \sin \theta),$$

which is called the *polar form* of z (since r and θ are the *polar-coordinates* of the point z).

EXERCISE (vii)
Express the following complex numbers in polar form

(a) -1,   (b) i,   (c) $-1 + i$,   (d) $\dfrac{1}{2} + \dfrac{i\sqrt{3}}{2}$.

EXERCISE (viii)
Find the locus of the points determined by the equations

(a) $|z - i| = 3$,   (b) $|z + i| + |z - i| = 4$,   (c) $\text{Im}(z^2) = 8$,
(d) $z\bar{z} = 4$,   (e) $\bar{z} = z + 3i$,   (f) $\arg(z - a) = \arg(b - a)$, where a and b are complex numbers.

1.3 De Moivre's theorem

If $z = r(\cos \theta + i \sin \theta)$ and $z' = r'(\cos \theta' + i \sin \theta')$,

then $zz' = rr'(\cos \theta + i \sin \theta)(\cos \theta' + i \sin \theta')$

$$= rr'(\cos \theta \cos \theta' - \sin \theta \sin \theta' + i\{\sin \theta \cos \theta' + \cos \theta \sin \theta'\})$$

$$= rr'(\cos \{\theta + \theta'\} + i \sin \{\theta + \theta'\}).$$

In other words (cf. Ex. (v)(a))

$$|zz'| = rr' = |z||z'|,$$

and

$$\arg(zz') = \arg(z) + \arg(z').$$

In general it is true that if $z_1 = r_1(\cos \theta_1 + i \sin \theta_1), \ldots,$
$z_n = r_n(\cos \theta_n + i \sin \theta_n)$, then

$$z_1 \ldots z_n = r_1 \ldots r_n(\cos(\theta_1 + \cdots + \theta_n) + i \sin(\theta_1 + \cdots + \theta_n)).$$

The proof of this theorem is by induction on n. The theorem is trivially true when $n = 1$, and we have already proved it for $n = 2$. Assume then the induction hypothesis, namely that

$$z_1 \ldots z_{n-1} = r_1 \ldots r_{n-1}(\cos(\theta_1 + \cdots + \theta_{n-1})$$
$$+ i \sin(\theta_1 + \cdots + \theta_{n-1})).$$

Then

$$z_1 \ldots z_{n-1} z_n = r_1 \ldots r_{n-1}(\cos(\theta_1 + \cdots + \theta_{n-1})$$
$$+ i \sin(\theta_1 + \cdots + \theta_{n-1}))r_n(\cos \theta_n + i \sin \theta_n)$$
$$= r_1 \ldots r_{n-1}r_n(\cos(\theta_1 + \cdots + \theta_{n-1} + \theta_n)$$
$$+ i \sin(\theta_1 + \cdots + \theta_{n-1} + \theta_n))$$

using the case when $n = 2$. Thus by assuming the validity of the theorem for the case $n - 1$, we have proved the theorem for the case n. By the principle of induction therefore, the theorem is true for all positive integral values of n. As a special case we have: if $z = r(\cos \theta + i \sin \theta)$ then for any positive integer n,

$$z^n = r^n(\cos n\theta + i \sin n\theta).$$

Since

$$r(\cos \theta + i \sin \theta) = \frac{r}{r'}(\cos \{\theta - \theta'\} + i \sin \{\theta - \theta'\}) . r'(\cos \theta' + i \sin \theta'),$$

then

$$\frac{z}{z'} = \frac{r}{r'}(\cos \{\theta - \theta'\} + i \sin \{\theta - \theta'\}),$$

provided $z' \neq 0$. Thus (cf. Ex. (v)(b))

$$\left| \frac{z}{z'} \right| = \frac{r}{r'} = \frac{|z|}{|z'|},$$

and

$$\arg \left(\frac{z}{z'} \right) = \arg z - \arg z'.$$

Therefore, for any positive integer n, and for $z \neq 0$

$$|z^{-n}| = \left| \frac{1}{z^n} \right| = \frac{1}{r^n} = r^{-n},$$

$$\arg (z^{-n}) = \arg \left(\frac{1}{z^n} \right) = - \arg (z^n) = -n(\arg z),$$

and so

$$z^{-n} = r^{-n}(\cos\{-n\theta\} + i \sin\{-n\theta\}).$$

We have proved *De Moivre's Theorem*:

If $z \neq 0$, and $z = r(\cos \theta + i \sin \theta)$, then for any integer n

$$z^n = r^n(\cos n\theta + i \sin n\theta).$$

A particular case of De Moivre's theorem is obtained if we take z to be a point on the *unit circle* (i.e., the circle in the Argand diagram of unit radius with centre at the origin), for then $r = 1$, and

$$(\cos \theta + i \sin \theta)^n = \cos n\theta + i \sin n\theta.$$

Example 1　Prove that

$$\cos 5\theta = 16 \cos^5 \theta - 20 \cos^3 \theta + 5 \cos \theta,$$
$$\sin 5\theta = \sin \theta \, (16 \cos^4 \theta - 12 \cos^2 \theta + 1).$$

Solution　By De Moivre's theorem, and then expanding by the binomial theorem, we have

$$\begin{aligned}
\cos 5\theta + i \sin 5\theta &= (\cos \theta + i \sin \theta)^5 \\
&= \cos^5 \theta + 5 \cos^4 \theta (i \sin \theta) + 10 \cos^3 \theta(i \sin \theta)^2 \\
&\quad + 10 \cos^2 \theta(i \sin \theta)^3 + 5 \cos \theta(i \sin \theta)^4 \\
&\quad + (i \sin \theta)^5 \\
&= (\cos^5 \theta - 10 \cos^3 \theta \sin^2 \theta + 5 \cos \theta \sin^4 \theta) \\
&\quad + i(5 \cos^4 \theta \sin \theta - 10 \cos^2 \theta \sin^3 \theta + \sin^5 \theta).
\end{aligned}$$

Equating real and imaginary parts, we obtain

$$\cos 5\theta = \cos^5 \theta - 10 \cos^3 \theta \sin^2 \theta + 5 \cos \theta \sin^4 \theta$$
$$= \cos^5 \theta - 10 \cos^3 \theta (1 - \cos^2 \theta) + 5 \cos \theta (1 - \cos^2 \theta)^2,$$
$$\sin 5\theta = 5 \cos^4 \theta \sin \theta - 10 \cos^2 \theta \sin^3 \theta + \sin^5 \theta$$
$$= \sin \theta \{5 \cos^4 \theta - 10 \cos^2 \theta (1 - \cos^2 \theta) + (1 - \cos^2 \theta)^2\}.$$

On simplification, these yield the desired results.

Example 2 Prove that $\cos^4 \theta = \frac{1}{8} (\cos 4\theta + 4 \cos 2\theta + 3)$.

Solution Put $z = \cos \theta + i \sin \theta$. Then $z^{-1} = \cos \theta - i \sin \theta$, and so $\cos \theta = \frac{1}{2} (z + z^{-1})$. Therefore

$$\cos^4 \theta = \frac{1}{16} (z + z^{-1})^4 = \frac{1}{16} (z^4 + 4z^2 + 6 + 4z^{-2} + z^{-4})$$
$$= \frac{1}{16} \{(z^4 + z^{-4}) + 4(z^2 + z^{-2}) + 6\}.$$

The proof is completed when we remark that, as a consequence of De Moivre's theorem, $z^4 + z^{-4} = 2 \cos 4\theta$ and $z^2 + z^{-2} = 2 \cos 2\theta$.

EXERCISE (ix)
Prove that

(a) $\cos 3\theta = 4 \cos^3 \theta - 3 \cos \theta$,
     $\sin 3\theta = 3 \sin \theta - 4 \sin^3 \theta$;
(b) $\cos 4\theta = 8 \cos^4 \theta - 8 \cos^2 \theta + 1$,
     $\sin 4\theta = \sin \theta (8 \cos^3 \theta - 4 \cos \theta)$;
(c) $8 \sin^4 \theta = \cos 4\theta - 4 \cos 2\theta + 3$;
(d) $16 \sin^5 \theta = \sin 5\theta - 5 \sin 3\theta + 10 \sin \theta$.

Example 3 Show that

$$\cos \theta + \cos 3\theta + \cos 5\theta + \cdots + \cos(2n - 1)\theta = \frac{\sin 2n\theta}{2 \sin \theta}.$$

Solution Let $z = \cos \theta + i \sin \theta$. By De Moivre's theorem, $z^{2n} = \cos 2n\theta + i \sin 2n\theta$ and $z^{-2n} = \cos 2n\theta - i \sin 2n\theta$. Therefore

$$2i \sin 2n\theta = z^{2n} - z^{-2n}$$
$$= (z - z^{-1})(z^{2n-1} + z^{2n-3} + \cdots + z + z^{-1} + \cdots + z^{-(2n-3)}$$
$$+ z^{-(2n-1)})$$
$$= (z - z^{-1})(\{z^{2n-1} + z^{-(2n-1)}\} + \{z^{2n-3} + z^{-(2n-3)}\}$$
$$+ \cdots + \{z + z^{-1}\})$$
$$= 2i \sin \theta (2 \cos(2n - 1)\theta + 2 \cos(2n - 3)\theta + \cdots + 2 \cos \theta).$$

EXERCISE (x)
Show that

(a) $\frac{1}{2} + \cos 2\theta + \cos 4\theta + \cos 6\theta + \cos 8\theta = \dfrac{\sin 9\theta}{2 \sin \theta}$,

(b) $\frac{1}{2} - \cos 2\theta + \cos 4\theta - \cos 6\theta + \cos 8\theta = \dfrac{\cos 9\theta}{2 \cos \theta}$.

1.4   Polynomial equations

Let $P(z)$ denote the polynomial

$$a_n z^n + a_{n-1} z^{n-1} + \cdots + a_1 z + a_0$$

where the coefficients $a_0, a_1, \ldots, a_n$ are complex numbers, and we suppose that $a_n \neq 0$. Then $P(z)$ is said to be of *degree n*. It is easily shown that the degree of a product of two polynomials is the sum of their respective degrees.

Any complex number α for which $P(\alpha) = 0$ is called a *zero* of the polynomial $P(z)$, or alternatively, a *root* of the polynomial equation $P(z) = 0$.

Note that, for any positive integer m,

$$z^m - \alpha^m = (z - \alpha)(z^{m-1} + \alpha z^{m-2} + \alpha^2 z^{m-3} + \cdots + \alpha^{m-2} z$$
$$+ \alpha^{m-1}). \qquad (1)$$

Thus if α is a zero of $P(z)$, then

$$P(z) = P(z) - P(\alpha)$$
$$= a_n(z^n - \alpha^n) + a_{n-1}(z^{n-1} - \alpha^{n-1}) + \cdots + a_1(z - \alpha).$$

It follows from (1) that every term in the right-hand side has $(z - \alpha)$ as a factor, and therefore $P(z)$ has $(z - \alpha)$ as a factor. Thus

$$P(z) = (z - \alpha)Q(z) \qquad (2)$$

where $Q(z)$ is some polynomial of degree $n - 1$.

If α' is a zero of $P(z)$ different from α, then

$$(\alpha' - \alpha)Q(\alpha') = P(\alpha') = 0,$$

and since $\alpha' - \alpha \neq 0$, then $Q(\alpha') = 0$ and so $Q(z)$ has $(z - \alpha')$ as a factor, i.e., $(z - \alpha)(z - \alpha')$ is a factor of $P(z)$. By induction we prove that
   if $\alpha_1, \alpha_2, \ldots, \alpha_m$ are distinct zeros of $P(z)$, then
   $$P(z) = (z - \alpha_1)(z - \alpha_2) \ldots (z - \alpha_m)R(z),$$

where $R(z)$ is a polynomial of degree $n - m$.

A polynomial of degree n cannot therefore have more than n distinct zeros. There is of course the question whether a polynomial need have any zeros at all. This is answered by the *Fundamental Theorem of*

Algebra (not proved in this book), which states that every polynomial has at least one zero. Thus the polynomial $Q(z)$ in (2) has a zero, and so has a linear factor. By continuing in this manner we see that *every polynomial of degree n has n (not necessarily distinct) zeros.* Let these n zeros be $\alpha_1, \alpha_2, \ldots, \alpha_n$. Then by repeating the argument given above we deduce that

$$P(z) = a_n(z - \alpha_1)(z - \alpha_2) \ldots (z - \alpha_n).$$

By multiplying out the right-hand side, this gives

$$P(z) = a_n\{z^n - (\alpha_1 + \alpha_2 + \cdots + \alpha_n)z^{n-1} + \cdots + (-1)^n\alpha_1\alpha_2 \ldots \alpha_n\},$$

from which we deduce the following formulae for the sum and the product of the zeros of $P(z)$:

$$\alpha_1 + \cdots + \alpha_n = -\frac{a_{n-1}}{a_n},$$

$$\alpha_1 \ldots \alpha_n = (-1)^n \frac{a_0}{a_n}.$$

Example 4   Show that

$$1 + \cos\frac{2\pi}{5} + \cos\frac{4\pi}{5} + \cos\frac{6\pi}{5} + \cos\frac{8\pi}{5} = 0,$$

$$\sin\frac{2\pi}{5} + \sin\frac{4\pi}{5} + \sin\frac{6\pi}{5} + \sin\frac{8\pi}{5} = 0.$$

Solution   The problem is equivalent to showing that

$$(1) + \left(\cos\frac{2\pi}{5} + i\sin\frac{2\pi}{5}\right) + \left(\cos\frac{4\pi}{5} + i\sin\frac{4\pi}{5}\right)$$

$$+ \left(\cos\frac{6\pi}{5} + i\sin\frac{6\pi}{5}\right) + \left(\cos\frac{8\pi}{5} + i\sin\frac{8\pi}{5}\right) = 0.$$

By De Moivre's theorem, each term in parenthesis is a zero of the polynomial $z^5 - 1$. Since we have five such terms, then these are precisely all the zeros of the polynomial. Their sum must be zero, since the coefficient of z^4 in the polynomial is 0.

EXERCISE (xi)
Find the roots of the equations

(a) $8z^3 - 6z = 1$ (Use Ex. (ix)(a)),   (b) $16z^5 - 20z^3 + 5z = 1$ (Use Example 1).

If the polynomial $P(z)$ has *real* coefficients, and if α is a zero of $P(z)$, then using the results of Ex. (iii) we have

$$
\begin{aligned}
0 &= \overline{a_n\alpha^n + a_{n-1}\alpha^{n-1} + \cdots + a_1\alpha + a_0} \\
&= \bar{a}_n\bar{\alpha}^n + \bar{a}_{n-1}\bar{\alpha}^{n-1} + \cdots + \bar{a}_1\bar{\alpha} + \bar{a}_0 \\
&= a_n\bar{\alpha}^n + a_{n-1}\bar{\alpha}^{n-1} + \cdots + a_1\bar{\alpha} + a_0.
\end{aligned}
$$

That is, the conjugate $\bar{\alpha}$ of α is also zero of $P(z)$. Thus, *the zeros of a polynomial with real coefficients occur in conjugate pairs.* A consequence of this is that *every polynomial with real coefficients can be factored into a product of linear and quadratic terms with real coefficients.*

The proof of this last statement is by induction on the degree n of $P(z)$. It is clearly true for $n = 2$. Our induction hypothesis is to assume it is true for any polynomial of degree less than n. Let $P(z)$ be a polynomial of degree n, and suppose that α is a zero of $P(z)$, $P(z) = (z - \alpha)Q(z)$. If α is real, then the coefficients of $Q(z)$ are also real, and the induction hypothesis applied to $Q(z)$ completes the proof for $P(z)$. If α is not real, then $\bar{\alpha} \neq \alpha$, and so α, $\bar{\alpha}$ are distinct zeros of $P(z)$. Therefore

$$
P(z) = (z - \alpha)(z - \bar{\alpha})R(z) = (z^2 - 2\,\mathrm{Re}\,(\alpha)z + |\alpha|^2)R(z).
$$

Since $z^2 - 2\,\mathrm{Re}\,(\alpha)z + |\alpha|^2$ has real coefficients, then so has $R(z)$, and once again the induction hypothesis applied to $R(z)$ completes the proof.

1.5   Roots of unity

The roots of the equation $z^n = 1$, where n is a positive integer, are called the *nth roots of unity*. Let $\omega = \cos(2\pi/n) + i\sin(2\pi/n)$. By De Moivre's theorem

$$
\omega^r = \cos\frac{2\pi r}{n} + i\sin\frac{2\pi r}{n} \quad \text{for } r = 1, 2, \ldots, n
$$

and

$$
(\omega^r)^n = \cos 2\pi r + i\sin 2\pi r = 1.
$$

The n complex numbers $\omega, \omega^2, \ldots, \omega^{n-1}, \omega^n (=1)$ are therefore all nth roots of unity. Since they are represented in the Argand plane by the vertices of a regular polygon of n sides inscribed in the unit circle, they are all distinct points. As there cannot be more than n nth roots of unity, we deduce that *the nth roots of unity are the complex numbers* $1, \omega, \omega^2, \ldots, \omega^{n-1}$. Since the coefficient of z^{n-1} in the equation is 0, then the sum of the roots is 0 (cf. Example 4), i.e.,

$$
1 + \omega + \omega^2 + \cdots + \omega^{n-1} = 0.
$$

EXERCISE (xii)
(a) Show that $(-1/2 + i(\sqrt{3}/2))^3 = 1$, and express the other cube roots of unity in terms of $-1/2 + i(\sqrt{3}/2)$.
(b) Find all the sixth roots of unity, and solve the equation $z^6 = (z + i)^6$.
(c) Solve the equation $(1 + z)^5 = (1 - z)^5$.
(d) If the roots of the equation $z^3 + 2iz^2 - z + 2i = 0$ are z_1, z_2, z_3, prove that

$$\frac{1}{z_1 + i} + \frac{1}{z_2 + i} + \frac{1}{z_3 + i} = 0.$$

To find the nth roots of any complex number α, i.e., the solutions of the equation $z^n = \alpha$, we first express α in polar form $\alpha = r(\cos \theta + i \sin \theta)$. By De Moivre's theorem the complex number $\sigma = r^{1/n}(\cos \theta/n + i \sin \theta/n)$ then certainly satisfies $z^n = \alpha$. We obtain n distinct solutions of the equation (and therefore all the solutions) if we multiply σ by each of the nth roots of unity:

$$\sigma, \sigma\omega, \sigma\omega^2, \ldots, \sigma\omega^{n-1}.$$

EXERCISE (xiii)
Solve

(a) $z^2 = -3 + 4i$, (b) $z^2 + z^{-2} = i$, (c) $z^{10} + z^5 + 1 = 0$,
(d) $z^9 + 1 = 0$.

CHAPTER 2

Sequences and Series

2.1 Limit of a sequence

A *sequence* is a set of (not necessarily distinct) real numbers s_n, one for each positive integer n, arranged in order of ascending n:

$$s_1, s_2, s_3, \ldots, s_n, \ldots \tag{1}$$

The number s_n is called the *nth term of the sequence.* Thus

$$1, \frac{1}{2}, \frac{1}{3}, \frac{1}{4}, \frac{1}{5}, \ldots, \tag{2}$$

and

$$0, 1, 0, 1, 0, \ldots \tag{3}$$

are examples of sequences. In (2), the fourth term is $\frac{1}{4}$, while the fourth term of (3) is 1.

The sequence (1) is usually denoted by $\{s_n\}$. In some cases we have a formula for s_n, and so, for instance, $\{1/n\}$ denotes the sequence (2), and $\{\frac{1}{2}(1 + (-1)^n)\}$ denotes the sequence (3).

EXERCISE (i)
Write down the first six terms of each of the following sequences:

(a) $\{n\}$,   (b) $\left\{\dfrac{n+1}{n}\right\}$,   (c) $\{(-1)^n\}$,   (d) $\{2^n\}$,

(e) $\{n + (-1)^n\}$.

In the study of sequences we are concerned mainly with the ultimate behaviour of the terms of the sequence. Thus, for instance, the terms of the sequence (2) get close to the value 0 for large values of n. We describe this situation by saying that the terms of the sequence (2) "tend to 0", and that "0 is the limit of the sequence (2)". Another example is the sequence

$$0, \frac{1}{2}, \frac{2}{3}, \frac{3}{4}, \frac{4}{5}, \ldots, \tag{4}$$

where $s_n = 1 - 1/n$. For large values of n the terms of (4) are very close to 1, and if one were to guess the limit of this sequence, one would undoubtedly give the answer as 1. Not all sequences have limits. The sequence (3) has terms which are alternately 0 and 1, and no matter how far along the sequence one goes, there is no number to which the terms of (3) get close. We say that the sequence (3) has no limit.

We have not yet in fact defined the limit of a sequence. Our discussion above was aimed at channelling our thoughts in the right direction. Phrases such as "close to 1" and "large values of n" are too imprecise, and may convey different meanings to different people, and may also depend on the context in which they are used. Our task then is to give an unambiguous and precise definition of limit.

A sequence $\{s_n\}$ is said to have limit s if, given any positive number ϵ, there exists an integer N (which depends on ϵ) such that

$$|s_n - s| < \epsilon \quad \text{for all } n > N.$$

Let us persuade ourselves of the reasonableness of this definition, by means of a diagram. Consider the diagram (Fig. 2) in which each term s_n of our sequence is represented by a dot at the point (n, s_n). Let S

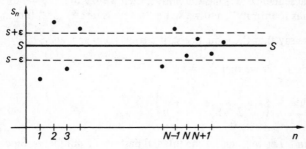

Figure 2

denote the line parallel to the horizontal axis at a distance s from it, and let the strip indicated by dotted lines consist of those points whose distance from S is less than ϵ. Then if s is the limit of the sequence $\{s_n\}$, by our definition there is a certain point N on the horizontal axis beyond which all the dots lie within the strip. Naturally, if we take a smaller ϵ, that is, if we take a narrower strip, then we must expect to have to go further to the right before all the dots lie within the strip. Thus, if we take a smaller ϵ, then in general the N required will be greater. The point of our definition, however, is that no matter how small we take ϵ there always is an N with the required property.

Example 1   Prove that the sequence $\{1/2^n\}$ has limit 0.

Solution   Given any $\epsilon > 0$, take N to be any integer greater than ϵ^{-1}. Then if $n > N$, we have $2^n > n > N > \epsilon^{-1} > 0$, and so $1/2^n < \epsilon$.

Example 2   Prove that the sequence $\{(3n - 1)/(4n^2 + 2)\}$ has limit 0.

Solution   We first note that

$$\frac{3n-1}{4n^2+2} < \frac{3n}{4n^2+2} < \frac{3n}{4n^2} < \frac{1}{n}.$$

Given $\epsilon > 0$, take N to be any integer greater than ϵ^{-1}. Then if $n > N$, we have $1/n < 1/N < \epsilon$, and therefore $(3n-1)/(4n^2+2) < \epsilon$.

EXERCISE (ii)
Prove that the following sequences all have limit 0.

(a) $\left\{\dfrac{(-1)^n}{n!}\right\}$,   (b) $\left\{\dfrac{\sin n}{n}\right\}$,   (c) $\left\{\dfrac{1-(-1)^n}{n}\right\}$,

(d) $\{\sqrt{n+1}-\sqrt{n}\}$,   (e) $\{r^n\}$ where $-1 < r < 1$,   (f) $\left\{\dfrac{2^n n!}{n^n}\right\}$.

If the sequence $\{s_n\}$ has the limit s, then we say that "s_n tends to s as n tends to infinity", and we write "$s_n \to s$ as $n \to \infty$" or "$\lim_{n\to\infty} s_n = s$".

Also, we say that the sequence s_n is a *convergent sequence*.

Example 3   Prove that

$$\lim_{n\to\infty} \frac{1}{n}\left(1 + \frac{1}{2} + \frac{1}{3} + \cdots + \frac{1}{n}\right) = 0.$$

Solution   Let $[\alpha]$ denote the integral part of the number α. Since

$$1 + \frac{1}{2} + \cdots + \frac{1}{[\sqrt{n}]} \leqslant [\sqrt{n}] \leqslant \sqrt{n},$$

and for $n > 1$

$$\frac{1}{[\sqrt{n}]+1} + \frac{i}{[\sqrt{n}]+2} + \cdots + \frac{1}{n} \leqslant \frac{n - [\sqrt{n}]}{[\sqrt{n}]+1} < \frac{n}{\sqrt{n}} = \sqrt{n},$$

then

$$\frac{1}{n}\left(1 + \frac{1}{2} + \cdots + \frac{1}{n}\right) < \frac{\sqrt{n}}{n} + \frac{1}{\sqrt{n}} = \frac{2}{\sqrt{n}}.$$

Thus, given $\epsilon > 0$, take N to be any integer greater than $4\epsilon^{-2}$. Then if $n > N$, we have $n > 4\epsilon^{-2}$ and so $2/\sqrt{n} < \epsilon$. This implies that

$$\frac{1}{n}\left(1 + \frac{1}{2} + \cdots + \frac{1}{n}\right) < \epsilon.$$

Example 4   Prove that $\lim_{n\to\infty} (3n-1)/(4n+2) = \frac{3}{4}$.

Solution First note that

$$\left| \frac{3n-1}{4n+2} - \frac{3}{4} \right| = \frac{5}{4(2n+1)} < \frac{2}{2n+1} < \frac{1}{n}.$$

Given any $\epsilon > 0$, take N to be any integer greater than ϵ^{-1}. Then for $n > N$, we have $n^{-1} < \epsilon$, and so

$$\left| \frac{3n-1}{4n+2} - \frac{3}{4} \right| < \epsilon \quad \text{whenever } n > N.$$

EXERCISE (iii)
Prove the following statements.

(a) $\displaystyle \lim_{n \to \infty} \frac{2n^2 + 5n}{3n^2 + 2n + 6} = \frac{2}{3}$,   (b) $\displaystyle \lim_{n \to \infty} \frac{2n^5 - 7n^3}{5n^7 - 6} = 0$,

(c) $\displaystyle \lim_{n \to \infty} \frac{1 + 3.5^n}{2 + 4.5^n} = \frac{3}{4}$.

EXERCISE (iv)
(a) Show that $\{s_n\}$ converges to s if both $\{s_{2n}\}$ and $\{s_{2n+1}\}$ converge to s.
(b) Show that if $\{s_n\}$ converges to s, then every *subsequence* of $\{s_n\}$ converges to s. ($\{a_n\}$ is called a *subsequence* of $\{s_n\}$ if, for each n, $a_n = s_{n'}$ for some n' and $n_1 < n_2$ implies $n_1' < n_2'$).

Sometimes, we can show that a sequence has a limit without actually knowing what that limit must be. A set S of real numbers is said to be *bounded above* if there is a number M such that $s \leqslant M$ for every s in S. M is then called an *upper bound* of S. Similarly, S is *bounded below* if there is an m such that $m \leqslant s$ for every s in S, and m is called a *lower bound* of S. We say simply that S is *bounded* if it is bounded both above and below, i.e., if $|s| \leqslant M$ for a fixed M, and for all s in S.

Before we state the theorem, we mention a property of real numbers: any bounded set of real numbers has a *least upper bound* and a *greatest lower bound*.

The sequence $\{s_n\}$ is *monotonic increasing* if $s_1 \leqslant s_2 \leqslant s_3 \leqslant \ldots$, and is called *monotonic decreasing* if $s_1 \geqslant s_2 \geqslant s_3 \geqslant \cdots$.

PROPOSITION 2.1.
Every monotonic increasing sequence of real numbers which is bounded above is convergent.

Proof
Suppose $\{s_n\}$ is monotonic increasing and bounded above. Then clearly it is bounded, and so has a least upper bound s, say. Thus $0 \leqslant s - s_n$ for all n. For every $\epsilon > 0$, we have $s - \epsilon < s$, and since s is the least upper bound of $\{s_n\}$, then $s - \epsilon$ is not an upper bound, i.e. there is an

integer N such that $s - \epsilon < s_N$. This implies that $s - \epsilon < s_n$ for all $n > N$, and so $0 \leqslant s - s_n < \epsilon$ for all $n > N$. This proves the convergence, and the limit of $\{s_n\}$ is s.

Example 5 If $e_n = (1 + 1/n)^n$, show that $\{e_n\}$ has a limit e, with $2 < e \leqslant 3$.

Solution Expand e_n by the binomial theorem:

$$e_n = 1 + \sum_{r=1}^{n} \binom{n}{r} \frac{1}{n^r},$$

where $\binom{n}{r}$ denotes the binomial coefficient $\dfrac{n!}{r!(n-r)!}$. Also,

$$e_{n+1} = 1 + \sum_{r=1}^{n+1} \binom{n+1}{r} \frac{1}{(n+1)^r}.$$

But

$$\binom{n}{r} \frac{1}{n^r} = \frac{1}{r!} \frac{n}{n} \frac{n-1}{n} \frac{n-2}{n} \cdots \frac{n-r+1}{n}$$

$$< \frac{1}{r!} \frac{n+1}{n+1} \frac{n}{n+1} \frac{n-1}{n+1} \cdots \frac{n-r+2}{n+1}$$

$$= \binom{n+1}{r} \frac{1}{(n+1)^r},$$

and therefore $e_n < e_{n+1}$ for $n = 1, 2, 3, \ldots$, i.e., $\{e_n\}$ is a monotonic increasing sequence, and $2 = e_1 < e_n$. Since

$$e_n < 1 + \sum_{r=1}^{n} \frac{1}{r!} \leqslant 1 + 1 + \sum_{r=2}^{n} \frac{1}{2^{r-1}} = 3 - \frac{1}{2^{n-1}} < 3,$$

then $\{e_n\}$ is bounded above. By Proposition 2.1, $\{e_n\}$ is convergent, with limit e, where e is the least upper bound of $\{e_n\}$. Since 3 is an upper bound, then $e \leqslant 3$.

2.2 Finding the limit

Quite clearly, not all sequences have limits. The sequence (3) is an example of a sequence which does not have a limit. Also the sequences (a), (c), (d) and (e) of Ex. (i) are all sequences without a limit. In the cases where limits do exist, however, it would be useful to devise techniques to find the limit, without having to go back to "first principles", in other words, without using the definition of a limit. First of all we prove

PROPOSITION 2.2
If a sequence has a limit, then this limit is unique.

Proof
Suppose s' and s'' are both limits of the sequence $\{s_n\}$. We must show that $s' = s''$. Since s' and s'' are limits, then given any $\epsilon > 0$, there exists N' such that $|s_n - s'| < \epsilon$ whenever $n > N'$, and there exists N'' such that $|s_n - s''| < \epsilon$ whenever $n > N''$. Put $N = \max\{N', N''\}$. Then $|s_n - s'| < \epsilon$ and $|s_n - s''| < \epsilon$ whenever $n > N$. But

$$|s'' - s'| = |(s_n - s') - (s_n - s'')| \leqslant |s_n - s'| + |s_n - s''|.$$

Since $|s_n - s'| + |s_n - s''| < 2\epsilon$ whenever $n > N$, then this implies that $|s'' - s'| < 2\epsilon$. Now $|s'' - s'|$ is a fixed number, while ϵ (and hence 2ϵ) is an arbitrary positive number, and therefore can be arbitrarily small. Thus $|s'' - s'| = 0$, and $s' = s''$.

Whenever we write $\lim\limits_{n \to \infty} s_n = s$, it will be implicit that we assume that the limit exists, and is s. Given two convergent sequences, we obtain new sequences by taking their sum, product and (if possible) their quotient. Our next task is to examine the convergence of these new sequences and to devise formulae for their limits.

PROPOSITION 2.3
If $\lim\limits_{n \to \infty} s_n = s$, *and if c is a constant, then* $\lim\limits_{n \to \infty} cs_n = cs$.

Proof
The case when $c = 0$ is trivial, so we may assume $c \neq 0$. Given any $\epsilon > 0$, consider the positive number $\epsilon/|c|$. Then there exists an integer N such that $|s_n - s| < \epsilon/|c|$ whenever $n > N$, i.e., $|cs_n - cs| < \epsilon$ whenever $n > N$.

The following proposition states that *the limit of a sum is the sum of the limits.*

PROPOSITION 2.4
If $\lim\limits_{n \to \infty} s_n = s$ *and* $\lim\limits_{n \to \infty} t_n = t$, *then* $\lim\limits_{n \to \infty} s_n + t_n = s + t$.

Proof
We observe that

$$|(s_n + t_n) - (s + t)| = |(s_n - s) + (t_n - t)| \leqslant |s_n - s| + |t_n - t|.$$

Consider $\epsilon > 0$. Then there exist integers N_1 and N_2 such that

$$|s_n - s| < \tfrac{1}{2}\epsilon \quad \text{whenever } n > N_1,$$

and

$$|t_n - t| < \tfrac{1}{2}\epsilon \quad \text{whenever } n > N_2.$$

If we take $N = \max\{N_1, N_2\}$, then

$$|s_n - s| + |t_n - t| < \tfrac{1}{2}\epsilon + \tfrac{1}{2}\epsilon = \epsilon$$

whenever $n > N$, and so

$$|(s_n + t_n) - (s + t)| < \epsilon \quad \text{whenever } n > N.$$

With Proposition 2.3, this gives us

COROLLARY 2.5
If c and d are constants, then $\lim\limits_{n \to \infty} cs_n + dt_n = cs + dt$.

A word of caution is appropriate here: the existence of $\lim\limits_{n \to \infty} s_n + t_n$ does not tell us anything about $\lim\limits_{n \to \infty} s_n$ or $\lim\limits_{n \to \infty} t_n$, which may in fact not exist at all. For example, with $s_n = n, t_n = -n$, then $s_n + t_n = 0$ and so $\lim\limits_{n \to \infty} s_n + t_n = 0$. However neither $\{n\}$ nor $\{-n\}$ have a limit.

PROPOSITION 2.6
Every convergent sequence is bounded.

Proof
Suppose that $\{s_n\}$ is convergent with limit s. Then there is an N such that $|s_n - s| < 1$ whenever $n > N$. Since

$$|s_n| = |s_n - s + s| \leqslant |s_n - s| + |s|,$$

then $|s_n| < 1 + |s|$ whenever $n > N$. If we take

$$M = \max\{|s_1|, |s_2|, \ldots, |s_N|, 1 + |s|\}, \text{ then } |s_n| \leqslant M \text{ for all } n.$$

The next proposition states that *the limit of a product is the product of the limits.*

PROPOSITION 2.7
If $\lim\limits_{n \to \infty} s_n = s$ *and* $\lim\limits_{n \to \infty} t_n = t$, *then* $\lim\limits_{n \to \infty} s_n t_n = st$.

Proof
By Proposition 2.6, we can find $M > 0$ such that $|s_n| < M$ for all n. Now

$$\begin{aligned} |s_n t_n - st| &= |s_n(t_n - t) + t(s_n - s)| \\ &\leqslant |s_n||t_n - t| + |t||s_n - s| \\ &\leqslant M|t_n - t| + (|t| + 1)|s_n - s|. \end{aligned}$$

Given $\epsilon > 0$, consider the two positive numbers $\epsilon/2M$ and $\epsilon/2(|t| + 1)$. Then there exist N_1 and N_2 such that

$$|s_n - s| < \epsilon/2(|t| + 1) \quad \text{whenever } n > N_1,$$

and

$$|t_n - t| < \epsilon/2M \quad \text{whenever } n > N_2.$$

If we take $N = \max\{N_1, N_2\}$, then using the inequality above we have

$$|s_n t_n - st| < \epsilon/2 + \epsilon/2 = \epsilon \quad \text{whenever } n > N.$$

The proposition which describes the limit of a quotient requires the following

LEMMA 2.8
If $\lim_{n \to \infty} t_n = t$, *and if* $t \neq 0$, *then there exists a number* N_2 *such that* $|t_n| > \frac{1}{2}|t|$ *whenever* $n > N_2$.

Proof
Since $\frac{1}{2}|t| > 0$, then there exists N_2 such that

$$|t_n - t| < \frac{1}{2}|t| \quad \text{whenever } n > N_2.$$

But

$$|t| = |t_n - t - t_n| \leqslant |t_n - t| + |t_n|,$$

and so

$$|t| < \frac{1}{2}|t| + |t_n| \quad \text{whenever } n > N_2,$$

and the result follows.

PROPOSITION 2.9
If $\lim_{n \to \infty} t_n = t$, *and if* $t \neq 0$, *then* $\lim_{n \to \infty} 1/t_n = 1/t$.

Proof
Given $\epsilon > 0$, consider the positive number $\frac{1}{2}|t|^2 \epsilon$. Then there exists N_1 such that

$$|t_n - t| < \frac{1}{2}|t|^2 \epsilon \quad \text{whenever } n > N_1.$$

From Lemma 2.8, there exists N_2 such that $|t_n| > \frac{1}{2}|t|$ whenever $n > N_2$. If we take $N = \max\{N_1, N_2\}$, then

$$\left| \frac{1}{t_n} - \frac{1}{t} \right| = \frac{|t_n - t|}{|t_n||t|} < \frac{\frac{1}{2}|t|^2 \epsilon}{\frac{1}{2}|t||t|} = \epsilon, \quad \text{whenever } n > N.$$

As a consequence of Propositions 2.7 and 2.9, we have

COROLLARY 2.10

If $\lim\limits_{n \to \infty} s_n = s$ *and* $\lim\limits_{n \to \infty} t_n = t$, *and if* $t_n \neq 0, t \neq 0$, *then* $\lim\limits_{n \to \infty} \dfrac{s_n}{t_n} = \dfrac{s}{t}$.

We conclude this list of useful propositions about limits with one which goes by the appropriate name of *The Sandwich Theorem*.

PROPOSITION 2.11

Suppose that $\{s_n\}$, $\{t_n\}$ *and* $\{u_n\}$ *are sequences of real numbers. If there is a positive integer* N_0 *such that* $s_n \leqslant t_n \leqslant u_n$ *for all* $n > N_0$, *and if* s_n *and* u_n *both converge to the limit* l, *then* $\{t_n\}$ *converges to* l.

Proof

Given $\epsilon > 0$, there exist N_1 and N_2 such that

$$l - \epsilon < s_n < l + \epsilon \quad \text{whenever } n > N_1$$

and

$$l - \epsilon < u_n < l + \epsilon \quad \text{whenever } n > N_2.$$

If we take $N = \max\{N_0, N_1, N_2\}$, then

$$l - \epsilon < s_n \leqslant t_n \leqslant u_n < l + \epsilon \quad \text{whenever } n > N.$$

Therefore $|t_n - l| < \epsilon$ whenever $n > N$.

Example 6 Find the limit of the sequence

$$\left\{ \frac{1 + n + n^2}{n(n - 1)(n - 2)} \right\}.$$

Solution We first rewrite the nth term of the sequence as

$$\left(\frac{1}{n^3} + \frac{1}{n^2} + \frac{1}{n} \right) \Big/ \left(1 - \frac{1}{n} \right) \left(1 - \frac{2}{n} \right).$$

Then the limit of the numerator is the sum of the limits of the functions $1/n^3$, $1/n^2$ and $1/n$. Since each of these limits is 0, then the limit of the numerator is 0. The limit of the denominator is the product

$$\lim \left(1 - \frac{1}{n} \right) \cdot \lim \left(1 - \frac{2}{n} \right),$$

which is equal to 1 since each of these limits is 1. Therefore we can deduce that the limit of our original sequence is $0/1$, i.e., is 0.

Example 7 Show that $a^n/n! \to 0$ as $n \to \infty$, for any fixed number a.

Solution We may as well assume that $a > 0$. Choose an integer N such that $N > 2a$. Then for large values of n (say, $n > N$)

$$\frac{a^n}{n!} = \frac{a}{1}\frac{a}{2}\cdots\frac{a}{N}\frac{a}{N+1}\cdots\frac{a}{n}$$

$$< \frac{a^N}{N!}(\tfrac{1}{2})^{n-N} = \frac{(2a)^N}{N!}(\tfrac{1}{2})^n.$$

Since $(\tfrac{1}{2})^n \to 0$ as $n \to \infty$ (Example 1), then $((2a)^N/N!)(\tfrac{1}{2})^n \to 0$ as $n \to \infty$. By the Sandwich Theorem therefore, $a^n/n! \to 0$ as $n \to \infty$.

Example 8 Prove that $(1 + n + n^2)^{1/n} \to 1$ as $n \to \infty$.

Solution Put $(1 + n + n^2)^{1/n} = 1 + u_n$. Then clearly $u_n > 0$, and we must show that $u_n \to 0$ as $n \to \infty$. Expanding by the binomial theorem, we have

$$1 + n + n^2 = (1 + u_n)^n$$

$$= 1 + nu_n + \frac{n(n-1)}{2!}u_n^2 + \frac{n(n-1)(n-2)}{3!}u_n^3 + \cdots + u_n^n$$

$$> \frac{n(n-1)(n-2)}{3!}u_n^3 \quad \text{for } n \geqslant 3.$$

If follows then that

$$0 < u_n^3 < \frac{3!(1 + n + n^2)}{n(n-1)(n-2)}.$$

From Example 6 and the Sandwich Theorem we deduce that $u_n^3 \to 0$, and so $u_n \to 0$ as $n \to \infty$.

EXERCISE (v)
Find the limits of the following sequences.

(a) $\left\{\left(\dfrac{2n-5}{3n+4}\right)^3\right\}$;   (b) $\left\{\dfrac{2n^2+3n}{2n^2+1}\right\}$;   (c) $\left\{\dfrac{n(n+2)}{n+1} - \dfrac{n^3}{n^2+1}\right\}$;

(d) $\{(1+n)^{1/n}\}$;   (e) $\left\{\dfrac{x^n}{n}\right\}$ when $0 < x \leqslant 1$; what happens if $x > 1$?

EXERCISE (vi)
If the sequence $\{s_n\}$ is defined by the equations $s_1 = 1$, $s_{n+1} = \sqrt{(12 + s_n)}$ for $n \geqslant 1$, prove that $\{s_n\}$ is monotonic increasing and bounded above. Show also that $\lim\limits_{n \to \infty} s_n = 4$.

EXERCISE (vii)

Let s_1 and s_2 be two real numbers with $s_1 < s_2$. Write $s_{n+2} = \frac{1}{2}(s_n + s_{n+1})$ for $n \geqslant 1$. Show that the sequence $\{s_n\}$ has a monotonic increasing subsequence $s_1, s_3, s_5, \ldots$, and a monotonic decreasing subsequence $s_2, s_4, s_6, \ldots$, and also that $\lim\limits_{n \to \infty} s_n$ exists and has the value $\frac{1}{3}(s_1 + 2s_2)$.

EXERCISE (viii)

Prove, by induction on r, that if $x \geqslant 1$ then

$$x^{r+1} - (x-1)^{r+1} \leqslant (r+1)x^r \leqslant (x+1)^{r+1} - x^{r+1}.$$

Deduce that

$$\frac{n^{r+1}}{r+1} \leqslant 1^r + 2^r + 3^r + \cdots + n^r \leqslant \frac{(n+1)^{r+1}}{r+1},$$

and hence prove that

(a) $\lim\limits_{n \to \infty} (1^r + 2^r + 3^r + \cdots + n^r)/n^{r+1} = \dfrac{1}{r+1}$;

(b) $\lim\limits_{n \to \infty} \dfrac{n(1^4 + 2^4 + 3^4 + \cdots + n^4)}{1^5 + 2^5 + 3^5 + \cdots + n^5} = \dfrac{6}{5}$.

2.3   Series

An expression

$$a_1 + a_2 + a_3 + \cdots + a_n + \cdots \tag{5}$$

which involves real numbers $a_1, a_2, a_3, \ldots, a_n, \ldots$, is called an *infinite series* (or briefly, a *series*). In giving this expression a name we are not claiming that this process of "infinite addition" necessarily has a meaningful interpretation. Our aim here is to consider circumstances under which such an interpretation can be given.

An immediate first step is to consider what happens if we add up just the first few terms of the series. We denote by s_n the *nth partial sum* of the series, viz.,

$$s_n = a_1 + a_2 + a_3 + \cdots + a_n.$$

This gives us a sequence $\{s_n\}$ of partial sums.

For example, the series

$$\frac{1}{2} + \frac{1}{2^2} + \frac{1}{2^3} + \cdots + \frac{1}{2^n} + \cdots \tag{6}$$

gives the sequence $\{1 - 1/2^n\}$ of partial sums, since

$$1 - \frac{1}{2^n} = \frac{1}{2} + \frac{1}{2^2} + \frac{1}{2^3} + \cdots + \frac{1}{2^n}.$$

The more terms of the series (6) we take, quite clearly, the closer the value of the partial sum gets to the value 1. More precisely, the sequence $\{1 - 1/2^n\}$ has the limit 1. We say then that the series (6) *converges*, and has the *sum* 1.

More generally then, the series (5) is said to *converge*, to the *sum s*, if its associated sequence $\{s_n\}$ of partial sums has the limit *s*. The series (5) is often denoted by $\Sigma_{n=1}^{\infty} a_n$, or Σa_n. In this case we write $\Sigma_{n=1}^{\infty} a_n = s$.

Of course, not all series are convergent series. The series

$$1 - 1 + 1 - 1 + \cdots + (-1)^{n+1} + \cdots \tag{7}$$

for example has partial sums which are alternately 1 and 0, and so does not converge. A series which does not converge is said to *diverge*.

Example 9   Show that the series $\Sigma_{n=1}^{\infty} 1/n(n + 1)$ is convergent, and find its sum.

Solution   Here, $a_n = 1/n(n + 1) = 1/n - 1/(n + 1)$, and so

$$s_n = \frac{1}{1.2} + \frac{1}{2.3} + \frac{1}{3.4} + \cdots + \frac{1}{n(n + 1)}$$

$$= \left(\frac{1}{1} - \frac{1}{2}\right) + \left(\frac{1}{2} - \frac{1}{3}\right) + \left(\frac{1}{3} - \frac{1}{4}\right) + \cdots + \left(\frac{1}{n} - \frac{1}{n + 1}\right)$$

$$= 1 - \frac{1}{n + 1}$$

Hence $\lim_{n \to \infty} s_n = 1$, and so $\Sigma_{n=1}^{\infty} 1/n(n + 1)$ converges, with sum 1.

EXERCISE (ix)

(a) If $|x| < 1$, show that the geometric series $\Sigma_{n=1}^{\infty} x^{n-1}$ converges with sum $1/(1 - x)$.
(b) Show that the series $\Sigma_{n=1}^{\infty} (2n + 1)/n^2(n + 1)^2$ converges, and find its sum.

It is of great importance in mathematics to be able to tell whether a particular series is convergent or divergent. The next proposition gives us a necessary condition for convergence.

PROPOSITION 2.12

If Σa_n converges, then $\lim_{n \to \infty} a_n = 0$.

Proof

Suppose Σa_n converges with sum s, i.e., suppose $s_n \to s$ as $n \to \infty$. Then given $\epsilon > 0$, there exists N such that $|s_n - s| < \frac{1}{2}\epsilon$ for all $n \geqslant N$. Therefore

$$|a_n| = |s_n - s_{n-1}|$$
$$= |(s_n - s) - (s_{n-1} - s)|$$
$$\leqslant |s_n - s| + |s_{n-1} - s|,$$

and so $|a_n| < \frac{1}{2}\epsilon + \frac{1}{2}\epsilon = \epsilon$ for all $n > N$.

The proposition is very useful as it permits us to deduce, by inspection alone, that certain series are not convergent. The series (7), for example, is not convergent since its nth term $(-1)^{n+1}$ does not tend to 0.

EXERCISE (x)

Show that the following series do not converge.

(a) $\sum_{n=1}^{\infty} \frac{2n}{3n+1}$;   (b) $\sum_{n=1}^{\infty} (\sqrt{n+1} - \sqrt{n})$;   (c) $\sum_{n=1}^{\infty} n^{1/n}$.

A very common mistake made by novices is to assume that the converse of this proposition holds. THE CONVERSE IS NOT TRUE. We give below an example of a series (the *harmonic series*) whose nth term tends to 0, yet which is divergent.

Example 10   Prove that the series $\Sigma_{n=1}^{\infty} 1/n$ is divergent.

Solution   Since the terms of the harmonic series are all positive, then the sequence $\{s_n\}$ of partial sums is monotonic increasing. Also,

$$s_{2^m} = 1 + \frac{1}{2} + \frac{1}{3} + \frac{1}{4} + \cdots + \frac{1}{2^m}$$

$$= 1 + \left(\frac{1}{2}\right) + \left(\frac{1}{3} + \frac{1}{4}\right) + \left(\frac{1}{5} + \frac{1}{6} + \frac{1}{7} + \frac{1}{8}\right) + \cdots + \left(\frac{1}{2^{m-1}+1}\right.$$

$$+ \frac{1}{2^{m-1}+2} + \cdots + \frac{1}{2^m}\left.\right).$$

Since each bracketed sum is greater than or equal to 1/2 (and we have m bracketed sums), then

$$s_{2^m} \geqslant 1 + \frac{1}{2}m > \frac{1}{2}m.$$

Thus, if $n > 2^m$, then $s_n > s_{2^m} > (1/2)m$. But given any positive number A, we can find $m > 2A$, and for all $n > 2^m$ we have $s_n > A$. Therefore $\{s_n\}$ is not convergent, and so $\Sigma 1/n$ is divergent.

Quite obviously, the propositions which we have proved for sequences will imply corresponding propositions for series.

PROPOSITION 2.13
If Σa_n converges to s, and c is a constant, then Σca_n converges to cs.

Proof
If s_n denotes the nth partial sum of Σa_n, then the nth partial sum of Σca_n is cs_n. The proposition then follows from Proposition 2.3.

PROPOSITION 2.14
If Σa_n and Σb_n converge with sums s and t respectively, then $\Sigma(a_n + b_n)$ converges to s + t.

Proof
This follows from Proposition 2.4.

The next proposition indicates the effect on convergence of removing the first few terms of a series. The proof is left as an exercise to the reader.

PROPOSITION 2.15
(i) *If $\Sigma_{n=1}^{\infty} a_n$ converges to s then $\Sigma_{n=N+1}^{\infty} a_n$ converges to $s - \Sigma_{n=1}^{N} a_n$.*
(ii) *if $\Sigma_{n=N+1}^{\infty} a_n$ converges to t, then $\Sigma_{n=1}^{\infty} a_n$ converges to $t + \Sigma_{n=1}^{N} a_n$.*

2.4 Tests for convergence

It would be tiresome indeed if, each time we wished to decide the convergence or divergence of a series, we had to return to the definitions. Often this is not practicable, for instance when the nth partial sum s_n is not expressible in a convenient form. It is useful therefore to devise some rules which tell us directly, without complicated calculations, whether a series is convergent.

For the present we restrict our attention to series of *non-negative terms*, i.e., series Σa_n for which $a_n \geqslant 0$. For such a series, the sequence $\{s_n\}$ of partial sums is monotonic increasing. If Σa_n is not convergent it follows from Proposition 2.1 that $\{s_n\}$ is not bounded. In other words, for every M, there exists an N such that $s_N > M$. In this case we write $s_n \to \infty$. ("s_n *tends to infinity*"). Clearly, $s_n \to \infty$ implies Σa_n is divergent.

Our first test compares our series with one which is already known to converge or diverge.

PROPOSITION 2.16 (The Comparison Test)

(i) *If $0 \leqslant a_n \leqslant b_n$ for all n, and if Σb_n is convergent, then Σa_n is convergent.*

(ii) *If $0 \leqslant b_n \leqslant a_n$ for all n, and if Σb_n is divergent, then Σa_n is divergent.*

Proof
Let $s_n = a_1 + a_2 + \cdots + a_n$, $t_n = b_1 + b_2 + \cdots + b_n$. Then both $\{s_n\}$ and $\{t_n\}$ are monotonic increasing.

(i) Since $t_n \to t$, say, then $t_n \leqslant t$. But $s_n \leqslant t_n$ for all n, and so $s_n \leqslant t$. By Proposition 2.1, Σa_n converges.
(ii) Since $t_n \leqslant s_n$, and $t_n \to \infty$, then $s_n \to \infty$ and Σa_n diverges.

Example 11   Show that $\Sigma_{n=1}^{\infty} 1/n^2$ is convergent.

Solution   We have already shown that $\Sigma\, 1/n(n+1)$ is convergent (Example 9), and hence $\Sigma\, 2/n(n+1)$ is convergent (Proposition 2.13). Since $0 < 1/n^2 < 2/n(n+1)$, then $\Sigma\, 1/n^2$ is convergent.

Example 12   Show that $\Sigma_{n=1}^{\infty} (n + \sqrt{n})/(2n^3 - 1)$ is convergent.

Solution   Since

$$0 < \frac{n + \sqrt{n}}{2n^3 - 1} \leqslant \frac{2n}{2n^3 - n} = \frac{2}{2n^2 - 1} \leqslant \frac{2}{n^2},$$

and (Example 11, Proposition 2.13) $\Sigma\, 2/n^2$ is convergent, then $\Sigma (n + \sqrt{n})/(2n^3 - 1)$ is convergent.

Example 13   Show that $\Sigma_{n=1}^{\infty} (2n + 3)/(n+1)(n+2)$ is divergent.

Solution   Since

$$\frac{2n + 3}{(n+1)(n+2)} > \frac{2(n+1)}{(n+1)(n+2)} = \frac{2}{n+2} > \frac{1}{2n},$$

and $\Sigma\, 1/2n$ is divergent (Example 10), then $\Sigma (2n + 3)/(n+1)(n+2)$ is divergent.

EXERCISE (xi)
Discuss the convergence of the following series.

(a) $\sum \dfrac{3}{4n^3 - 3}$;   (b) $\sum \dfrac{5n}{2n^2 + 2n + 1}$;   (c) $\sum \left(\dfrac{n+1}{2n}\right)^n$;

(d) $\sum \dfrac{1}{\sqrt{n}}$;   (e) $\sum \dfrac{2^n}{5^{n-1}}$; $\sum \dfrac{n+1}{(n+2)\sqrt{n+3}}$.

In view of Proposition 2.15, we can obtain a slight improvement of the comparison test. If there is an N such that $0 \leqslant a_n \leqslant b_n$ for all

$n > N$, then the convergence of $\Sigma_{n=N+1}^{\infty} b_n$ will imply the convergence of $\Sigma_{n=N+1}^{\infty} a_n$, and hence of $\Sigma_{n=1}^{\infty} a_n$. On the other hand, if $0 \leqslant b_n \leqslant a_n$ for all $n > N$, then the divergence of $\Sigma_{n=N+1}^{\infty} b_n$ will imply the divergence of $\Sigma_{n=N+1}^{\infty} a_n$, and hence of $\Sigma_{n=1}^{\infty} a_n$. Note that these conclusions hold even if some of the first N terms of either series are not non-negative. We have then

COROLLARY 2.17 (The Comparison Test)
(i) *If $0 \leqslant a_n \leqslant b_n$ for all $n > N$, then the convergence of Σb_n implies the convergence of Σa_n.*
(ii) *If $0 \leqslant b_n \leqslant a_n$ for all $n > N$, then the divergence of Σb_n implies the divergence of Σa_n.*

EXERCISE (xii)
Examine for convergence the series

(a) $\displaystyle\sum_{n=1}^{\infty} \frac{n^2 - 100n}{n^4}$;   (b) $\displaystyle\sum_{n=1}^{\infty} \frac{n^2 - 8n}{2n^3 - 3}$;   (c) $\displaystyle\sum_{n=1}^{\infty} \frac{4\sqrt{n} - 3\sqrt{(n+1)}}{n+2}$.

The technique of the comparison test, naturally, is to compare the series with others whose behaviour is already known. The drawback here is that the test is not, in a sense, self-contained. The *ratio test* which follows removes this defect and examines a series without recourse to other known series.

PROPOSITION 2.18 (D'Alembert's ratio test).
If Σa_n is a series of positive terms and if $\lim\limits_{n \to \infty} (a_{n+1}/a_n)$ exists and has the value l, then

(i) *if $l < 1$, Σa_n converges;*
(ii) *if $l > 1$, Σa_n diverges.*

Proof
(i) If $l < 1$, choose r so that $l < r < 1$. Since $r - l > 0$ (then by taking $\epsilon = r - l$ in the definition of limit) there exists N such that $a_{n+1}/a_n < l + (r - l) = r$ for all $n > N$. Therefore $a_{N+2} < a_{N+1}r$, $a_{N+3} < a_{N+2}r < a_{N+1}r^2$, and in general, $a_n < a_{N+1}r^{-(N+1)}r^n$ for all $n > N$. Thus, by the comparison test (since the geometric series Σr^n converges) Σa_n converges.
(ii) If $l > 1$, (take $\epsilon = l - 1$ in the definition of limit) there exists N such that $a_{n+1}/a_n > l - (l - 1) = 1$ for all $n > N$, i.e., $a_{n+1} > a_n$ for all $n > N$. Thus $n > N$ implies $a_n > a_N$, and so a_n does not tend to zero. By Proposition 2.12, Σa_n diverges.

It is worth stressing that *the ratio test tells us nothing about the convergence of a series if $l = 1$.* Indeed, we have seen examples (Examples 10 and 11) where $l = 1$ but in one case the series diverges while in the other case the series converges. So if $l = 1$, the ratio test is ineffective and some other test is required. A common mistake in the use of the ratio test is to forget that it is the *limit of the ratio* a_{n+1}/a_n which concerns us. The fact that $a_{n+1}/a_n < 1$ for all n is not sufficient reason for Σa_n to converge. In Example 10 again, we have $a_{n+1}/a_n = n/(n + 1) < 1$, yet that series diverges.

Example 14   Show that $\Sigma n!/n^n$ converges.

Solution   Since

$$\frac{a_{n+1}}{a_n} = \frac{(n + 1)!}{(n + 1)^{n+1}} \cdot \frac{n^n}{n!} = \left(\frac{n}{n + 1}\right)^n,$$

then (Example 5) $\lim_{n \to \infty} a_{n+1}/a_n = 1/e < 1/2 < 1$. The series therefore converges by the ratio test. (Alternatively, observe that $n!/n^n \leqslant 2/n^2$ and use the comparison test.)

EXERCISE (xiii)
Discuss the convergence of the series whose nth terms are given by

(a) $\dfrac{n^2}{2^n}$;   (b) $\dfrac{n^4}{3^{n^2}}$;   (c) $\dfrac{(n!)^3}{(3n)!}$;   (d) $4^n \dfrac{n!}{n^n}$.

EXERCISE (xiv)
Find for what positive values of x the following series are convergent.

(a) $\Sigma (5x)^n$;   (b) $\sum \dfrac{x^n}{n}$;   (c) $\sum \dfrac{1}{n(1 + x^2)^n}$;   (d) $\Sigma (nx)^n$;

(e) $\Sigma x^{n!}$.

It is already clear that $\Sigma\, 1/n$ and $\Sigma\, 1/n^2$ are useful series to have in our repertoire of series whose behaviour is known. If α is a rational number, the comparison test tells us that $\Sigma\, 1/n^\alpha$ converges for $\alpha \geqslant 2$, and diverges for $\alpha \leqslant 1$. The no-man's land between 1 and 2 is covered by

Example 15   Show that, if α is a rational number, then $\Sigma\, 1/n^\alpha$ diverges for $\alpha \leqslant 1$, and converges for $\alpha > 1$.

Solution   We need only show convergence for $\alpha > 1$. The sequence $\{s_n\}$ of nth partial sums is monotonic increasing and so, in view of

Proposition 2.1 it is sufficient to show that $\{s_n\}$ is bounded above. Now

$$s_{2^m} = 1 + \left(\frac{1}{2^\alpha} + \frac{1}{3^\alpha}\right) + \left(\frac{1}{4^\alpha} + \frac{1}{5^\alpha} + \frac{1}{6^\alpha} + \frac{1}{7^\alpha}\right) + \cdots$$

$$+ \left(\frac{1}{(2^{m-1})^\alpha} + \frac{1}{(2^{m-1}+1)^\alpha} + \cdots + \frac{1}{(2^m-1)^\alpha}\right) + \frac{1}{(2^m)^\alpha}$$

$$< 1 + \left(\frac{1}{2^\alpha} + \frac{1}{2^\alpha}\right) + \left(\frac{1}{4^\alpha} + \frac{1}{4^\alpha} + \frac{1}{4^\alpha} + \frac{1}{4^\alpha}\right) + \cdots$$

$$+ \left(\frac{1}{(2^{m-1})^\alpha} + \frac{1}{(2^{m-1})^\alpha} + \cdots + \frac{1}{(2^{m-1})^\alpha}\right) + \frac{1}{(2^m)^{\alpha-1}}$$

$$= 1 + \frac{1}{2^{\alpha-1}} + \frac{1}{4^{\alpha-1}} + \cdots + \frac{1}{(2^{m-1})^{\alpha-1}} + \frac{1}{(2^m)^{\alpha-1}}$$

$$= \frac{1 - \left(\dfrac{1}{2^{\alpha-1}}\right)^{m+1}}{1 - \dfrac{1}{2^{\alpha-1}}} < \left(1 - \frac{1}{2^{\alpha-1}}\right)^{-1}.$$

Given any positive integer n we can find m such that $n \leqslant 2^m$, and so $s_n < s_{2^m}$, i.e., $s_n < (1 - 1/2^{\alpha-1})^{-1}$ for all n.

2.5 Alternating series

There are certain series which are easily dealt with even though they contain terms of mixed sign. These are the *alternating series*, where the terms are alternately positive and negative:

$$a_1 - a_2 + a_3 - a_4 + \cdots + (-1)^{n+1}a_n + \cdots$$

or

$$-a_1 + a_2 - a_3 + a_4 - \cdots + (-1)^n a_n + \cdots \qquad (8)$$

with $a_1, a_2, \ldots, a_n, \ldots$ positive.

PROPOSITION 2.19 **(Leibniz's test)**
An alternating series (8), *where* $a_n \geqslant a_{n+1}$ *and* a_n *tends to zero, is a convergent series.*

Proof
Since the two series of (8) differ only by a factor of -1, we need only consider one of them. Consider the partial sum

$$s_{2n} = (a_1 - a_2) + (a_3 - a_4) + \cdots + (a_{2n-1} - a_{2n})$$
$$= a_1 - (a_2 - a_3) - \cdots - (a_{2n-2} - a_{2n-1}) - a_{2n}.$$

Since each bracket is non-negative, then

$$0 \leqslant s_{2(n-1)} \leqslant s_{2n} \leqslant a_1.$$

Thus (Proposition 2.1) s_{2n} has a limit s, say, and $s \leqslant a_1$. Also $s_{2n+1} = s_{2n} + a_{2n+1}$, and since $s_{2n} \to s$, $a_{2n+1} \to 0$, then $s_{2n+1} \to s$. It follows then (Ex. (iv)) that $s_n \to s$ as $n \to \infty$.

EXERCISE (xv)
Show that the following series converge.

(a) $\Sigma(-1)^{n+1} \dfrac{1}{n^\alpha}$ for rational $\alpha > 0$;   (b) $\Sigma(-1)^{n+1} \dfrac{1}{n}\left(1 + \tfrac{1}{2} + \cdots + \dfrac{1}{n}\right)$

(Use Example 3);   (c) $\Sigma(-1)^{n+1} \dfrac{1}{2n + (-1)^n}$;

(d) $\Sigma(-1)^{n+1} \left(\sin \dfrac{1}{\sqrt{n}}\right) \dfrac{1}{2n - 1}$.

Example 16   Given that $\Sigma_{n=1}^{\infty} 1/n^2$ converges to $\pi^2/6$ show that

$$\sum_{n=1}^{\infty} \frac{(-1)^{n+1}}{n^2} = \frac{\pi^2}{12}.$$

Solution   Let $s_n = 1/1^2 + 1/2^2 + 1/3^2 + \cdots + 1/n^2$, and $t_{2n} = 1/1^2 - 1/2^2 + 1/3^2 - \cdots - 1/(2n)^2$.
Then

$$t_{2n} = s_{2n} - 2\left(\frac{1}{2^2} + \frac{1}{4^2} + \cdots + \frac{1}{(2n)^2}\right)$$

$$= s_{2n} - \frac{1}{2}s_n.$$

Since $s_{2n} \to (\pi^2/6)$ and $s_n \to (\pi^2/6)$, then $t_{2n} \to (\pi^2/12)$, and so $t_n \to (\pi^2/12)$.

2.6   Absolute convergence

When a series Σa_n contains an infinite number of terms of each sign, and if it is not an alternating series, then none of the tests so far discussed apply. One method is to consider the series $\Sigma|a_n|$ whose terms are all non-negative, and on which therefore our previous tests may be used. The next proposition gives the relation between $\Sigma|a_n|$ and Σa_n.

PROPOSITION 2.20
If $\Sigma|a_n|$ converges then so does Σa_n.

Proof
Define $a_n^+ = a_n$ if $a_n \geq 0$, $a_n^+ = 0$ if $a_n \leq 0$, and $a_n^- = 0$ if $a_n \geq 0$, $a_n^- = a_n$ if $a_n \leq 0$. Since $0 \leq a_n^+ \leq |a_n|$, and $0 \leq -a_n^- \leq |a_n|$ for all n, then by the comparison test Σa_n^+ and $\Sigma(-a_n^-)$ (and hence, Σa_n^-) are convergent. But $a_n = a_n^+ + a_n^-$, and therefore Σa_n converges (Proposition 2.14).

The series Σa_n is said to be *absolutely convergent* if the series $\Sigma|a_n|$ is convergent. Our proposition therefore states that *an absolutely convergent series is convergent.*

Not all convergent series are absolutely convergent. The *alternating harmonic series* $\Sigma(-1)^{n+1} 1/n$ is a case in point. Such series Σa_n are called *conditionally convergent series*: Σa_n is convergent but $\Sigma|a_n|$ is not.

We recall that the terms of a series Σa_n are given in a definite order, this order being indicated by the suffix n of a_n. If we rearrange the terms of the series then there is no guarantee that the resulting series will converge even if the original series did. In fact, if Σa_n is conditionally convergent then it can be rearranged to converge to any sum we desire, or even to diverge (this is known as *Riemann's theorem*). When Σa_n is absolutely convergent, however, the situation is less drastic.

LEMMA 2.21
If Σa_n is a convergent series of non-negative terms and Σb_n is a rearrangement of Σa_n, then Σb_n converges and has the same sum as Σa_n.

Proof
Let s_n, t_n denote the nth partial sums of $\Sigma a_n, \Sigma b_n$ respectively. Suppose Σa_n converges to A. Each of the terms $b_1, \ldots, b_n$ is an a_i for some i. If we take m to be the largest value of i which occurs (naturally, m will depend on n) then $t_n \leq s_m$ since the set $b_1, \ldots, b_n$ is contained in the set $a_1, \ldots, a_m$. This implies $t_n \leq A$ since $s_m \leq A$, and therefore $\{t_n\}$ converges to B, say, where $B \leq A$. Finally, we observe that, if Σb_n is a rearrangement of Σa_n, then equally Σa_n is a rearrangement of Σb_n. By the argument above $A \leq B$, and this proves that $A = B$.

PROPOSITION 2.22
If Σb_n is a rearrangement of the absolutely convergent series Σa_n, then Σb_n converges absolutely also and has the same sum as Σa_n.

Proof
In the notation of Proposition 2.20, we have $a_n = a_n^+ + a_n^-$, $b_n = b_n^+ + b_n^-$, and Σa_n^+, Σa_n^- are convergent. Now Σb_n^+ is a rearrangement of Σa_n^+, which is a series of non-negative terms. By the lemma then, $\Sigma b_n^+ = \Sigma a_n^+$. Similarly, $\Sigma b_n^- = \Sigma a_n^-$, and so $\Sigma b_n = \Sigma a_n$.

EXERCISE (xvi)
Discuss the convergence and absolute convergence of:

(a) $\Sigma(\sqrt{(n+1)} - \sqrt{n})x^n$;   (b) $\Sigma(2^n + 3^n)x^n$;   (c) $\sum \dfrac{x^n}{n!}$;

(d) $\Sigma n! x^n$;   (e) $1 + \dfrac{1}{3^2} + \dfrac{1}{5^2} - \dfrac{1}{2^2} + \dfrac{1}{7^2} + \dfrac{1}{9^2} + \dfrac{1}{11^2} - \dfrac{1}{4^2} +++ - \cdots$.

2.7 Power series

A *power series in* x is a series

$$a_0 + a_1 x + a_2 x^2 + \cdots + a_n x^n + \cdots \tag{9}$$

where the $a_0, a_1, \ldots, a_n, \ldots$ are constants, and we regard x as a variable. Note that the numbering starts from $n = 0$. We denote the power series (9) by $\Sigma_{n=0}^{\infty} a_n x^n$ or simply $\Sigma a_n x^n$.

Clearly, the power series (9) converges when $x = 0$. We have examples where absolute convergence occurs (i) only for $x = 0$ (Ex. (xvi)(d)), (ii) for all values of x (Ex. (xvi)(c)), and (iii) for $|x| < R$ for some number R and divergence for $|x| > R$ (Ex. (xvi)(a), (b)). In fact, for power series, these are the only situations which can occur. We state this in the next proposition. Our treatment of power series throughout will be limited to statements of propositions whose proofs are, by and large, outside the scope of this book. We mark these propositions with a (*).

*PROPOSITION 2.23
For every power series in* x, *just one of the following possibilities must occur*:

  (i) *it converges only when* $x = 0$;
 (ii) *it converges absolutely for all* x;
(iii) *there is a number* R *such that the power series converges absolutely when* $|x| < R$ *and diverges when* $|x| > R$.

The number R is called the *radius of convergence* of the series. The cases (i) and (ii) above correspond to $R = 0$ and $R = \infty$.

Inside its radius of convergence a power series $\Sigma a_n x^n$ defines a function $f(x)$ of x, where $f(x) = \Sigma a_n x^n$. Note that $f(x)$ is not defined for $|x| > R$. The notions of continuity and differentiability are discussed in the next chapter, but we anticipate these ideas by stating

*PROPOSITION 2.24

(i) *The series $\Sigma_{n=0}^{\infty} a_n x^n$, $\Sigma_{n=0}^{\infty} n a_n x^{n-1}$ and $\Sigma_{n=0}^{\infty} (a_n/(n+1)) x^{n+1}$ all have the same radius of convergence R.*
(ii) *The function $f(x) = \Sigma a_n x^n$ is continuous and has a derivative $f'(x)$ for $|x| < R$, and $f'(x) = \Sigma n a_n x^{n-1}$.*

We can add and multiply two power series together to obtain another. This process and their properties are set out in

*PROPOSITION 2.25
Suppose that $\Sigma_{n=0}^{\infty} a_n x^n$ converges to $f(x)$ for $|x| < R_1$, and $\Sigma_{n=0}^{\infty} b_n x^n$ converges to $g(x)$ for $|x| < R_2$. Then*

(i) *$\Sigma_{n=0}^{\infty} (a_n + b_n) x^n$ converges absolutely for $|x| < \min\{R_1, R_2\}$ to the sum $f(x) + g(x)$,*
(ii) *$a_0 b_0 + (a_0 b_1 + a_1 b_0) x + (a_0 b_2 + a_1 b_1 + a_2 b_0) x^2 + \cdots$ converges absolutely for $|x| < \min\{R_1, R_2\}$ to the sum $f(x)g(x)$.*

The product of two series defined in (ii) is called the *Cauchy product.* The student may have noticed the similarity between this definition and the process of multiplying two polynomials in x and collecting together the powers of x. This Cauchy product of two power series (with the same value of x) is in fact a special case of a more general theorem about the product of (not necessarily power) series. The general theorem is

*PROPOSITION 2.26
Suppose $\Sigma_{n=0}^{\infty} a_n$ and $\Sigma_{n=0}^{\infty} b_n$ both converge absolutely to A and B respectively. Then the series $\Sigma_{n=0}^{\infty} c_n$, where*

$$c_n = a_0 b_n + a_1 b_{n-1} + \cdots + a_{n-1} b_1 + a_n b_0,$$

is absolutely convergent with sum AB.

While we will not use this fact, it may be of interest to state the following stronger version of Proposition 2.26.

*PROPOSITION 2.27
If Σa_n is absolutely convergent and Σb_n is convergent, then Σc_n is convergent with sum AB.*
The series Σc_n is called the *Cauchy product* of Σa_n and Σb_n.

2.8   The exponential and logarithmic functions

In this section we again anticipate the notions of continuity and differentiability which are discussed in the next chapter. The reader may, if he prefers, postpone the reading of this section until after Chapter 3. Those who are (even vaguely) familiar with differentiation will do himself no great harm by reading this section now.

The ratio test tells us that the series

$$1 + x + \frac{x^2}{2!} + \frac{x^3}{3!} + \cdots + \frac{x^n}{n!} + \cdots \tag{10}$$

converges absolutely for all values of x. The function defined by (10) is denoted by $E(x)$, and we observe that $E(x)$ is defined for all values of x. Differentiating term by term (Proposition 2.24 (ii)), we get

$$E'(x) = E(x) \tag{11}$$

for all x (cf. Chapter 3, Example 10). $E(x)$ is called the *exponential function*. It is one of the basic functions in mathematics, and we now examine the properties of this function in some detail.

PROPOSITION 2.28
$E(x_1)E(x_2) = E(x_1 + x_2)$ *for all* x_1, x_2.

Proof
Since $\Sigma_{n=0}^\infty x_1^n/n!$ and $\Sigma_{n=0}^\infty x_2^n/n!$ converge absolutely for all x_1, x_2, then by Proposition 2.26 their Cauchy product converges to $E(x_1)E(x_2)$. In other words

$$E(x_1)E(x_2) = c_0 + c_1 + c_2 + \cdots + c_n + \cdots$$

where

$$c_n = \frac{x_1^n}{n!} + \frac{x_1^{n-1}}{(n-1)!}x_2 + \frac{x_1^{n-2}}{(n-2)!}\frac{x_2^2}{2!} + \cdots + \frac{x_1^{n-r}x_2^r}{(n-r)!r!} + \cdots + \frac{x_2^n}{n!}$$

$$= \frac{1}{n!} \sum_{r=0}^n \binom{n}{r} x_1^{n-r} x_2^r$$

$$= \frac{(x_1 + x_2)^n}{n!} \quad \text{by the Binomial theorem.}$$

This gives us the required equality.
Since $E(0) = 1$, then $E(x)E(-x) = 1$ and so

COROLLARY 2.29

$$E(-x) = \frac{1}{E(x)} \text{ for all } x.$$

Clearly, from Proposition 2.28, we have

$$(E(x))^n = E(nx)$$

for any positive integer n. Thus, if p and q are positive integers then

$$\left(E\left(\frac{p}{q}\right)\right)^q = E(p) = (E(1))^p,$$

and so

$$E\left(\frac{p}{q}\right) = (E(1))^{p/q}.$$

This means that

$$E(x) = (E(1))^x \tag{12}$$

whenever x is a positive rational number; from Corollary 2.29, (12) holds for *any* rational number x. The function $E(x) - (E(1))^x$ is a continuous function of x which takes the value 0 whenever x is rational. It is a property of continuous functions that this must then take the value 0 for all real x (rational or not). This means that (12) holds for all real x. We denote the value $E(1)$ by the symbol e, i.e.,

$$e = 1 + 1 + \frac{1}{2!} + \frac{1}{3!} + \cdots + \frac{1}{n!} + \cdots,$$

and so the function $E(x)$ may be expressed in the form e^x. (The number e is irrational, does not satisfy any polynomial equation with integer coefficients, and to six significant figures has value 2.71828.)

We must admit that, so far we have not defined the function $a^x (a > 0)$ for irrational values of x. We remedy this by making the following definition: *if $a \geq 1$ and x is irrational, then the set $\{a^r$, such that r is rational and $r < x\}$ is bounded above, and a^x is defined to be the least upper bound of this set; if $0 < a < 1$ and x is irrational, then the set $\{a^r$, such that r is rational and $r < x\}$ is bounded below, and a^x is defined to be the greatest lower bound of this set.*

Now return to our main business. From (10) it follows that

$$e^x > 1 + x > 1 \quad \text{if } x > 0 \tag{13}$$

and *a fortiori*, $e^x > 0$ if $x > 0$. In view of Corollary 2.29, we have $e^x > 0$ *for all x.* This implies that $e^x > 1 + x$ for $x \leq -1$. For $-1 < x < 0$, the alternating series

$$\frac{x^2}{2!} + \frac{x^3}{3!} + \cdots + \frac{x^n}{n!} + \cdots$$

converges to a positive sum (see proof of Leibniz's test), and so $e^x - (1 + x) > 0$. We have shown then that

$$e^x > 1 + x \quad \text{for all } x \neq 0, \tag{14}$$

and

$$e^x \to \infty \quad \text{as } x \to \infty.$$

If $x_1 > x_2$, then $x_1 - x_2 > 0$ and $e^{x_1-x_2} > 1$ (from (13)), and so

$$e^{x_1} > e^{x_2} \quad if \ x_1 > x_2 \tag{15}$$

i.e., e^x *never assumes the same value for two distinct values of x.*

Let y be any positive number, and take α, β (possibly negative) such that $y < 1 + \beta$ and $1/y < 1 - \alpha$. Then from (14) we have $e^\alpha < y < e^\beta$, and from (15) we see that $\alpha < \beta$. The Intermediate value theorem for continuous functions tells us that the function e^x must take the value y for some x between α and β; and so, *every positive number is a value of the function e^x*, or, *the function e^x assumes every positive value just once.*

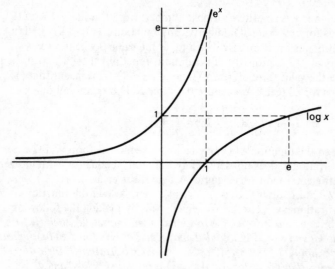

Figure 3

This statement enables us to define a new function $\log x$. For $x > 0$ there is a unique number, denoted by $\log x$, for which

$$x = e^{\log x}. \tag{16}$$

Immediately, we have

$$\log(1) = 0 \text{ and } \log(e) = 1. \tag{17}$$

From (14), we see that

$$\log x < x - 1 \quad if \ x \neq 1, x > 0,$$
$$i.e., \quad \log(1 + x) < x \quad if \ x \neq 0, x > -1. \tag{18}$$

Since

$$e^{\log x_1 x_2} = x_1 x_2 = e^{\log x_1} e^{\log x_2} = e^{\log x_1 + \log x_2},$$

then

$$\log x_1 x_2 = \log x_1 + \log x_2 \quad \text{if } x_1, x_2 > 0. \tag{19}$$

Combining this with (17) $\left(\text{take } x_1 = x, x_2 = \dfrac{1}{x}\right)$ we obtain

$$\log \frac{1}{x} = -\log x \quad \text{for } x > 0. \tag{20}$$

The function $\log x$ is not bounded, for if $\log x < A$ for all x then from (15) $x = e^{\log x} < e^A$ for all x, which is patently not true. Thus $\log x \to \infty$ as $x \to \infty$.

Put $y = \log x$. Then from (16) $x = e^y$, and

$$1 = \frac{d}{dx} e^y = e^y \frac{dy}{dx} = x \frac{dy}{dx},$$

and so

$$\frac{d}{dx} (\log x) = \frac{1}{x}. \tag{21}$$

Since $\log(1) = 0$, this gives us the alternative definition of $\log x$,

$$\log x = \int_1^x \frac{1}{t} \, dt \quad \text{for } x > 0. \tag{22}$$

EXERCISE (xvii)
Show that $\log x$ is a monotonic increasing function which takes every real value exactly once.

EXERCISE (xviii)
Use the inequality (18) to show that

$$\frac{1}{m+1} < \log(m+1) - \log m < \frac{1}{m} \quad \text{for } m > 0,$$

and deduce that, if $s_n = 1 + \dfrac{1}{2} + \dfrac{1}{3} + \cdots + \dfrac{1}{n}$, then

(a) $\log(n+1) < s_n < 1 + \log n$;
(b) the sequence $\{s_n - \log n\}$ tends to a limit (Euler's constant, denoted by γ).

By considering the expression

$$s_{2n} - s_n = (s_{2n} - \log 2n) - (s_n - \log n) + \log 2,$$

or otherwise, show that the series

$$1 - \frac{1}{2} + \frac{1}{3} - \frac{1}{4} + \cdots + \frac{(-1)^{n+1}}{n} + \cdots$$

converges to log 2.

EXERCISE (xix)

Show that the series

$$1 - \frac{1}{2} - \frac{1}{4} + \frac{1}{3} - \frac{1}{6} - \frac{1}{8} + \frac{1}{5} - \frac{1}{10} - \frac{1}{12} + \frac{1}{7} - \cdots$$

converges to $\frac{1}{2}$ log 2.

2.9 Hyperbolic functions

We define

$$\sinh x = \tfrac{1}{2}(e^x - e^{-x}), \qquad \cosh x = \tfrac{1}{2}(e^x + e^{-x}). \tag{23}$$

These are called the *hyperbolic sine* and *hyperbolic cosine* of x. In view of (10) and Proposition 2.25, they have the following representation as the sum of series:

$$\sinh x = x + \frac{x^3}{3!} + \frac{x^5}{5!} + \cdots + \frac{x^{2n+1}}{(2n+1)!} + \cdots, \tag{24}$$

$$\cosh x = 1 + \frac{x^2}{2!} + \frac{x^4}{4!} + \cdots + \frac{x^{2n}}{(2n)!} + \cdots, \tag{25}$$

which converge absolutely for all x.

From (23) it follows that

$$\frac{d}{dx} \sinh x = \cosh x \qquad \frac{d}{dx} \cosh x = \sinh x.$$

The similarity with the circular functions sin and cos is shown in

EXERCISE (xx)

Show that the following equalities hold for all x, y:

(a) $\cosh^2 x - \sinh^2 x = 1$;

(b) $\sinh(x + y) = \sinh x \cosh y + \cosh x \sinh y$;

(c) $\cosh(x + y) = \cosh x \cosh y + \sinh x \sinh y.$

By analogy with circular functions we define

$$\tanh x = \frac{\sinh x}{\cosh x}, \qquad \operatorname{sech} x = \frac{1}{\cosh x},$$

$$\coth x = \frac{\cosh x}{\sinh x}, \qquad \operatorname{cosech} x = \frac{1}{\sinh x}.$$

From (25) we see that $\cosh x \geqslant 1$ for all x, and so $\tanh x$, $\operatorname{sech} x$ are defined for all x (compare with $\tan x$, $\sec x$). Since $\sinh x = 0$ only when $x = 0$, then $\coth x$, $\operatorname{cosech} x$ are defined for all $x, x \neq 0$.

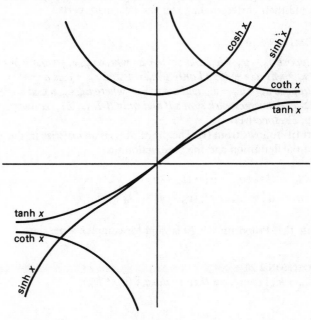

Figure 4

EXERCISE (xxi)

Prove from the definitions that

(a) $\dfrac{d}{dx} \tanh x = \operatorname{sech}^2 x = 1 - \tanh^2 x$;

(b) $\dfrac{d}{dx} \coth x = -\operatorname{cosech}^2 x \quad \text{if } x \neq 0$;

(c) $\dfrac{d}{dx} \operatorname{sech} x = -\tanh x \operatorname{sech} x$;

(d) $\dfrac{d}{dx} \operatorname{cosech} x = -\coth x \operatorname{cosech} x \quad \text{if } x \neq 0$.

2.10 Series of complex terms

So far the sequences and series under consideration had terms which were assumed to be real numbers. In fact closer examination of the definitions of convergence show that they make perfectly good sense even if the sequence $\{s_n\}$ of §2.1 and the series (5) of §2.3 were assumed to be complex. In fact, the student should make a point of checking that Propositions 2.2–2.10 (and their proofs) continue to be valid for sequences of complex numbers, while Propositions 2.12–2.15 (and their proofs) continue to be valid for series of complex numbers.

The following relationships hold between complex sequences and series and their corresponding real and imaginary parts.

PROPOSITION 2.30

(i) *Suppose $s_n = u_n + iv_n$ and $s = u + iv$, where u_n, v_n, u and v are real. Then $s_n \to s$ as $n \to \infty$ if and only if $u_n \to u$ and $v_n \to v$ as $n \to \infty$.*
(ii) *Suppose $a_n = u_n + iv_n$ and $s = u + iv$, where u_n, v_n, u and v are real. Then Σa_n converges with sum s if and only if Σu_n, Σv_n converge with sums u, v respectively.*

Part (ii) follows from (i). The proof of (i) is an exercise in the application of the definition and the observation that

$$|s_n - s| \leqslant |u_n - u| + |v_n - v|,$$
$$|u_n - u| \leqslant |s_n - s|, |v_n - v| \leqslant |s_n - s|.$$

We note that Proposition 2.20 is valid for complex series, i.e.,

PROPOSITION 2.20′
If $\Sigma |u_n + iv_n|$ converges then so does $\Sigma (u_n + iv_n)$.

Proof
Since $|u_n| \leqslant |u_n + iv_n|$, then by the comparison test $\Sigma |u_n|$ converges, and hence Σu_n converges. Similarly, Σv_n converges and therefore (Proposition 2.30 (ii)) $\Sigma (u_n + iv_n)$ converges.

This proposition therefore states that an absolutely convergent series (of complex terms) is necessarily convergent.

In §2.7 the treatment places no special restriction on a_n nor on x to be real. Thus the power series $\Sigma a_n x^n$ can in general be regarded as a series of complex terms. Propositions 2.23–2.27 continue to be valid in this case. An interesting and useful generalisation thus obtained is the function $E(z)$ defined by the series

$$1 + z + \frac{z^2}{2!} + \frac{z^3}{3!} + \cdots + \frac{z^n}{n!} + \cdots \tag{26}$$

for a complex number z. The series (26) converges absolutely for all values of z to a value denoted by $E(z)$, and from the analogue of Proposition 2.28 we have

$$E(x + iy) = E(x)E(iy).$$

For x, y real therefore this becomes

$$E(x + iy) = e^x E(iy).$$

Now

$$E(iy) = 1 + (iy) + \frac{(iy)^2}{2!} + \frac{(iy)^3}{3!} + \cdots + \frac{(iy)^n}{n!} + \cdots$$

$$= 1 + iy - \frac{y^2}{2!} - \frac{iy^3}{3!} + \frac{y^4}{4!} + \frac{iy^5}{5!} - \cdots$$

$$= \left(1 - \frac{y^2}{2!} + \frac{y^4}{4!} - \frac{y^6}{6!} + \cdots\right) + i\left(y - \frac{y^3}{3!} + \frac{y^5}{5!} - \frac{y^7}{7!} + \cdots\right).$$

EXERCISE (xxii)
Show that the (complex) series

$$1 - \frac{y^2}{2!} + \frac{y^4}{4!} - \frac{y^6}{6!} + \cdots \tag{27}$$

$$y - \frac{y^3}{3!} + \frac{y^5}{5!} - \frac{y^7}{7!} + \cdots \tag{28}$$

converge absolutely for all values of y.

The convergent series (27) therefore defines a function of y which we shall denote by Cos y. Similarly, the series (28) defines a function which we denote by Sin y. By abuse of notation we shall write e^z for the function $E(z)$. We have then

$$e^{iy} = \text{Cos } y + i \text{ Sin } y. \tag{29}$$

At this stage the student should distinguish between the functions Cos, Sin, and the trigometrically defined cos, sin (to begin with, cos and sin are not defined if y is complex but Cos and Sin are). We shall show later on (Chapter 3) that, if y is real, then indeed Cos y = cos y and Sin y = sin y. The student will recall that every complex number z may be represented in the form $z = r(\cos \theta + i \sin \theta)$. In view of (29) therefore we have $z = re^{i\theta}$. This is known as *Euler's form* for the complex number z.

EXERCISE (xxiii)

(a) Show that

$$\text{Cos } y = \tfrac{1}{2}(e^{iy} + e^{-iy}),$$

$$\text{Sin } y = \frac{1}{2i}(e^{iy} - e^{-iy}).$$

(b) By differentiating term by term, show that

$$\frac{d}{dx}\text{Cos } x = -\text{Sin } x, \qquad \frac{d}{dx}\text{Sin } x = \text{Cos } x.$$

(c) Use part (a) and Proposition 2.28 to show that

$$\text{Cos}^2 x + \text{Sin}^2 x = 1,$$

$$\text{Sin}(x + y) = \text{Sin } x \text{ Cos } y + \text{Cos } x \text{ Sin } y,$$

$$\text{Cos}(x + y) = \text{Cos } x \text{ Cos } y - \text{Sin } x \text{ Sin } y.$$

CHAPTER 3

Continuous Functions of One Variable

We assume throughout this book that the reader has had some experience with functions and with some of the more elementary techniques of calculus. In particular, it is assumed that he is familiar with the process of differentiation and can obtain the derivatives of simple functions. Nevertheless, familiarity with techniques often hide a somewhat vague idea of the underlying subject matter. In this chapter we give a brief account of continuity and differentiation, and obtain some basic results.

3.1 Limit of a function

Consider a function $f(x)$ which is defined in some interval, say $a \leqslant x \leqslant b$, and let $a < \alpha < b$. We are interested in the behaviour of the function "as x gets close to α". If there is a number l such that "$f(x)$ gets us close as we please to l provided x is taken sufficiently close to α", then intuitively we feel justified in thinking of l as the limit of $f(x)$ as x tends to α. In other words, $|f(x) - l|$ can be made as small as we like, provided we take $|x - \alpha|$ sufficiently small.

Straightaway we face the same objection as we had before concerning the limit of a sequence. The phrases "as close as we please", "sufficiently close" and "small" are not mathematically precise enough. As in the case of sequences we make a formal definition of limit to overcome these objections.

The function $f(x)$ is said to have limit l as x tends to α if, given any $\epsilon > 0$ there exists a δ (which depends on the choice of ϵ) such that $|f(x) - l| < \epsilon$ for all x satisfying $0 < |x - \alpha| < \delta$.

We write $\lim_{x \to \alpha} f(x) = l$, or $f(x) \to l$ as $x \to \alpha$.

Visually, this means that, given a strip of width 2ϵ symmetrically placed about the line $y = l$ (in Fig. 5) the value of $f(x)$ will lie within this strip provided the distance of x from α is less than δ and $x \neq \alpha$. It is important to note that in this definition we make no assumptions whatever about $f(\alpha)$. The value of $f(\alpha)$ may, possibly, lie outside the strip. In fact (since $f(\alpha)$ does not appear in the definition) it was not necessary even to suppose that the function $f(x)$ was defined at $x = \alpha$.

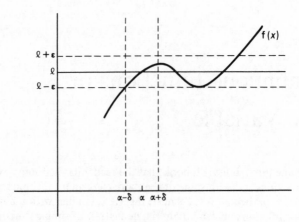

Figure 5

Consider, for example, the function $f(x)$ defined for $-1 \leqslant x \leqslant 1$ by

$$f(x) = \begin{cases} 1 + x & \text{for} \quad -1 \leqslant x < 0, \\ 1 - x & \text{for} \quad 0 < x \leqslant 1, \\ 0 & \text{for} \quad x = 0. \end{cases} \tag{1}$$

It is soon checked that $\lim\limits_{x \to 0} f(x) = 1$, while of course $f(0) = 0$.

Example 1   The function $f(x)$ is defined to be $(x^3 - 1)/(x - 1)$ for $x \neq 1$, and is not defined for $x = 1$. Show that $\lim\limits_{x \to 1} f(x) = 3$.

Solution   If $x \neq 1$, then

$$\left| \frac{x^3 - 1}{x - 1} - 3 \right| = |x^2 + x - 2| = |x - 1| \, |x + 2|.$$

Given $\epsilon > 0$, we note that $|x - 1| < 1$ implies $|x + 2| < 4$, and $|x - 1| < \frac{1}{4}\epsilon$ implies $4|x - 1| < \epsilon$. If we take $\delta = \min\{1, \frac{1}{4}\epsilon\}$, then $0 < |x - 1| < \delta$ implies $|x - 1| \, |x + 2| < 4|x - 1| < \epsilon$.

EXERCISE (i)
Show that

(a) $\lim\limits_{x \to 0} \dfrac{1 - x^2}{1 - \frac{1}{2}x^2} = 1$,   (b) $\lim\limits_{x \to 1} \dfrac{x^3 + 2x^2 - x - 2}{x^2 + 4x - 5} = 1$,

(c) $\lim\limits_{x \to 1} \dfrac{x + 1}{x^2 + 3x} = \frac{1}{2}$.

We give now some simple results concerning limits.

PROPOSITION 3.1

(i) *A function $f(x)$ cannot have more than one limit as $x \to \alpha$.*

(ii) *If $f(x) \to l$ as $x \to \alpha$, then there exists δ such that $f(x)$ is bounded for $0 < |x - \alpha| < \delta$.*

(iii) *If $f(x) \to l$ and $g(x) \to m$ as $x \to \alpha$, and if for some δ, $f(x) \leqslant g(x)$ whenever $0 < |x - \alpha| < \delta$, then $l \leqslant m$.*

(iv) *If $f(x) \to l$ as $x \to \alpha$, and for some δ, $f(x) \leqslant c$ (or $f(x) \geqslant c$) whenever $0 < |x - \alpha| < \delta$, then $l \leqslant c$ (or $l \geqslant c$).*

(v) *If $f(x) \to l$ and $g(x) \to l$ as $x \to \alpha$, and if for some δ, $f(x) \leqslant h(x) \leqslant g(x)$ whenever $0 < |x - \alpha| < \delta$, then $h(x) \to l$ as $x \to \alpha$ (Sandwich theorem).*

Proof

(i) Suppose $f(x) \to l_1$ and $f(x) \to l_2$ as $x \to \alpha$. Given $\epsilon > 0$ there exist δ_1, δ_2 such that

$$|f(x) - l_1| < \epsilon \quad \text{for all } x \text{ satisfying } 0 < |x - \alpha| < \delta_1,$$
$$|f(x) - l_2| < \epsilon \quad \text{for all } x \text{ satisfying } 0 < |x - \alpha| < \delta_2.$$

Thus

$$|l_1 - l_2| = |(f(x) - l_2) - (f(x) - l_1)| \leqslant |f(x) - l_2| + |f(x) - l_1|,$$

and so $|l_1 - l_2| < 2\epsilon$ whenever x satisfies $0 < |x - \alpha| < \delta = \min\{\delta_1, \delta_2\}$. But $|l_1 - l_2|$ is a constant, while ϵ is an arbitrary positive number, and so $|l_1 - l_2| = 0$, i.e., $l_1 = l_2$.

(ii) Take $\epsilon = 1$. There exists a δ such that

$$|f(x) - l| < 1 \quad \text{for all } x \text{ satisfying } 0 < |x - \alpha| < \delta.$$

For such x therefore, $|f(x)| - |l| < 1$, i.e.,

$$|f(x)| < |l| + 1.$$

(iii) Suppose $l > m$. Then $\frac{1}{2}(l - m) > 0$, and there exist δ_1, δ_2 such that

$$|f(x) - l| < \tfrac{1}{2}(l - m) \quad \text{for all } x \text{ satisfying } 0 < |x - \alpha| < \delta_1,$$
$$|g(x) - m| < \tfrac{1}{2}(l - m) \quad \text{for all } x \text{ satisfying } 0 < |x - \alpha| < \delta_2.$$

In particular, for all x satisfying $0 < |x - \alpha| < \min\{\delta_1, \delta_2, \delta\}$, we have

$$l - \tfrac{1}{2}(l - m) < f(x),$$

and

$$g(x) < m + \tfrac{1}{2}(l - m),$$

which implies $g(x) < f(x)$ for such x. This contradicts our hypothesis that $f(x) \leqslant g(x)$.

(iv) This is a corollary of (iii), obtained by taking $g(x)$ to be the constant function c.

(v) Given $\epsilon \to 0$, there exist δ_1, δ_2 such that

$$l - \epsilon < f(x) < l + \epsilon \qquad \text{whenever } 0 < |x - \alpha| < \delta_1,$$
$$l - \epsilon < g(x) < l + \epsilon \qquad \text{whenever } 0 < |x - \alpha| < \delta_2.$$

Therefore we have

$$l - \epsilon < f(x) \leqslant h(x) \leqslant g(x) < l + \epsilon \quad \text{whenever } 0 < |x - \alpha|$$
$$< \min\{\delta, \delta_1, \delta_2\}$$

i.e.,

$$|h(x) - l| < \epsilon \quad \text{for all } x \text{ satisfying } 0 < |x - \alpha| < \min\{\delta, \delta_1, \delta_2\}.$$

Example 2   Show that $\lim\limits_{x \to 0} \dfrac{\operatorname{Sin} x}{x} = 1$.

Solution   From the definition of $\operatorname{Sin} x$ (Chapter 2, (28)) we see that

$$\frac{\operatorname{Sin} x}{x} = 1 - \frac{x^2}{3!} + \frac{x^4}{5!} - \frac{x^6}{7!} + \cdots.$$

As we are interested only in the behaviour of this function near $x = 0$, we may as well restrict our consideration to those x with $|x| < 1$. Term by term comparison with the convergent geometric series yields

$$1 - \frac{x^2}{2} - \frac{x^4}{4} - \frac{x^6}{8} - \cdots < \frac{\operatorname{Sin} x}{x} < 1 + \frac{x^2}{2} + \frac{x^4}{4} + \frac{x^6}{8} + \cdots,$$

i.e.,

$$\frac{1 - x^2}{1 - \frac{1}{2}x^2} < \frac{\operatorname{Sin} x}{x} < \frac{1}{1 - \frac{1}{2}x^2}.$$

Since

$$\lim_{x \to 0} \frac{1 - x^2}{1 - \frac{1}{2}x^2} = 1 = \lim_{x \to 0} \frac{1}{1 - \frac{1}{2}x^2},$$

then the Sandwich theorem yields the desired result.

EXERCISE (ii)
Show that

(a) $\lim\limits_{x \to 0} \operatorname{Sin} x = 0$, (b) $\lim\limits_{x \to 0} \operatorname{Cos} x = 1$, (c) $\lim\limits_{x \to 0} \dfrac{1 - \operatorname{Cos} x}{x^2} = \frac{1}{2}$,

(d) $\lim\limits_{x \to 0} e^x - 1 = 0$, (e) $\lim\limits_{x \to 0} \dfrac{e^x - 1}{x} = 1$,

(f) $\lim\limits_{x \to 0} \dfrac{e^x - 1 - x}{x^2} = \frac{1}{2}$.

3.2   Finding the limit

The following proposition concerning limits allows us to calculate the limit of a function in very many cases without recourse to the basic definition. It is reminiscent of the propositions of §2.2.

PROPOSITION 3.2
Suppose that $f(x) \rightarrow l$ and $g(x) \rightarrow m$ as $x \rightarrow \alpha$. Then as $x \rightarrow \alpha$

(i) $cf(x) \rightarrow cl$ *for any constant c,*
(ii) $f(x) + g(x) \rightarrow l + m$,
(iii) $f(x)g(x) \rightarrow lm$,
(iv) $f(x)/g(x) \rightarrow l/m$ *provided $m \neq 0$.*

Proof

(i) is a consequence of (iii) if we take $g(x)$ to be the constant function $g(x) = c(=m)$.

(ii) Given $\epsilon > 0$, there exist δ_1, δ_2 such that

$$|f(x) - l| < \tfrac{1}{2}\epsilon \quad \text{for all } x \text{ satisfying } 0 < |x - \alpha| < \delta_1,$$

$$|g(x) - m| < \tfrac{1}{2}\epsilon \quad \text{for all } x \text{ satisfying } 0 < |x - \alpha| < \delta_2.$$

Hence for all x for which $0 < |x - \alpha| < \delta = \min\{\delta_1, \delta_2\}$, we have

$$|f(x) + g(x) - (l + m)| \leqslant |f(x) - l| + |g(x) - m| < \epsilon.$$

(iii) By Proposition 3.1 (ii), $f(x)$ and $g(x)$ are bounded in some region $0 < |x - \alpha| < \delta : |f(x)| < K, |g(x)| < K$, say. Then in particular, $|m| \leqslant K$. Given $\epsilon > 0$, there exist δ_1, δ_2 such that

$$|f(x) - l| < \frac{\epsilon}{2K} \quad \text{for all } x \text{ satisfying } 0 < |x - \alpha| < \delta_1,$$

$$|g(x) - m| < \frac{\epsilon}{2K} \quad \text{for all } x \text{ satisfying } 0 < |x - \alpha| < \delta_2.$$

Thus for $0 < |x - \alpha| < \min\{\delta, \delta_1, \delta_2\}$ we have

$$|f(x)g(x) - lm| = |f(x)(g(x) - m) + m(f(x) - l)|$$
$$\leqslant |f(x)||g(x) - m| + |m||f(x) - l|$$
$$< K\frac{\epsilon}{2K} + K\frac{\epsilon}{2K} = \epsilon.$$

(iv) It is sufficient to show that $1/g(x) \rightarrow 1/m$, as this with (iii) will imply (iv). Since $m \neq 0$, then $\tfrac{1}{2}|m| > 0$, and there exists δ such that

$$|g(x) - m| < \tfrac{1}{2}|m| \quad \text{for all } x \text{ satisfying } 0 < |x - \alpha| < \delta.$$

For such x therefore

$$|g(x)| > \tfrac{1}{2}|m|,$$

and in particular, $g(x) \neq 0$ so that $1/g(x)$ is defined. Now given $\epsilon > 0$, there exists δ' such that

$$|g(x) - m| < \tfrac{1}{2}m^2 \epsilon \quad \text{for all } x \text{ satisfying } 0 < |x - \alpha| < \delta'.$$

Hence, for $0 < |x - \alpha| < \min\{\delta, \delta'\}$ we have

$$\left|\frac{1}{g(x)} - \frac{1}{m}\right| = \left|\frac{g(x) - m}{g(x)m}\right| < \frac{|g(x) - m|}{\tfrac{1}{2}m^2} < \epsilon.$$

Example 3   Find $\lim\limits_{x \to 1} (x^3 + 2x^2 - x - 2)/(x^2 + x - 2)$.

Solution   Notice that we cannot use Proposition 3.2 (iv) directly as $\lim\limits_{x \to 1} x^2 + x - 2 = 0$. However, for $x \neq 1$, we have

$$\frac{x^3 + 2x^2 - x - 2}{x^2 + x - 2} = \frac{x^2 + 3x + 2}{x + 2}.$$

In evaluating $\lim\limits_{x \to 0}$, we are concerned with values of x *except* $x = 1$, and so

$$\lim_{x \to 1} \frac{x^3 + 2x^2 - x - 2}{x^2 + x - 2} = \lim_{x \to 1} \frac{x^2 + 3x + 2}{x + 2} = \frac{6}{3} = 2.$$

The function $f(x)$ is said to "tend to the limit l as x tends to infinity" if the value of $f(x)$ gets close to l for large enough values of x. Formally, we say

$f(x)$ has the limit l as x tends to infinity if, given any $\epsilon > 0$ there exists a number X (which depends on ϵ) such that

$$|f(x) - l| < \epsilon \quad \text{for all } x > X.$$

We write $\lim\limits_{x \to \infty} f(x) = l$, or "$f(x) \to l$ as $x \to \infty$".

The student may wish to prove for himself (as an exercise) that Proposition 3.2 is valid with α replaced by ∞.

Example 4   Evaluate $\lim\limits_{x \to \infty} (6x^3 + 2x^2 + 77)/(3x^3 + 1000)$.

Solution

$$\frac{6x^3 + 2x^2 + 77}{3x^3 + 1000} = \frac{6 + \dfrac{2}{x} + \dfrac{77}{x^3}}{3 + \dfrac{1000}{x^3}} \quad \text{for } x \neq 0.$$

Now $\lim\limits_{x \to \infty} 1/x^n = 0$ for $n = 1, 2, 3, \ldots$, and so $\lim\limits_{x \to \infty} (6 + 2/x + 77/x^3) = 6$, $\lim\limits_{x \to \infty} (3 + 1000/x) = 3$. Hence $\lim\limits_{x \to \infty} (6x^3 + 2x^2 + 77)/(3x^3 + 1000) = 6/3 = 2$.

For more complicated functions the following proposition is useful.

PROPOSITION 3.3
If $f(x) \to l$ as $x \to \alpha$ (but $f(x) \neq l$), and if $g(t) \to m$ as $t \to l$, then $g(f(x)) \to m$ as $x \to \alpha$.

Proof
Given $\epsilon > 0$, there exists $\delta_1 > 0$ such that

$$|g(t) - m| < \epsilon \quad \text{for all } t \text{ satisfying } 0 < |t - l| < \delta_1.$$

Associated with this δ_1 is a δ_2 such that

$$0 < |f(x) - l| < \delta_1 \quad \text{for all } x \text{ satisfying } 0 < |x - \alpha| < \delta_2.$$

Thus $0 < |x - \alpha| < \delta_2$ implies $|g(f(x)) - m| < \epsilon$.
(Note that we have only used the fact that $f(x) \neq l$ for values of x near α, $x \neq \alpha$.)

Example 5 Evaluate $\lim\limits_{x \to 1} (1 - \text{Cos}(x^2 - 1))/(1 - \text{Cos}^2(x - 1))$.

Solution

$$\frac{1 - \text{Cos}(x^2 - 1)}{1 - \text{Cos}^2(x - 1)} = \frac{1 - \text{Cos}(x^2 - 1)}{(x^2 - 1)^2} \cdot \frac{(x - 1)^2}{1 - \text{Cos}(x - 1)} \cdot \frac{(x + 1)^2}{1 + \text{Cos}(x - 1)} .$$

Since $x^2 - 1 \to 0$ as $x \to 1$, and $(1 - \text{Cos } t)/t^2 \to \frac{1}{2}$ as $t \to 0$, then

$$\lim\limits_{x \to 1} \frac{1 - \text{Cos}(x^2 - 1)}{(x^2 - 1)^2} = \frac{1}{2}. \quad \text{Similarly, } \lim\limits_{x \to 1} \frac{1 - \text{Cos}(x - 1)}{(x - 1)^2} = \frac{1}{2}.$$

Also $\lim\limits_{x \to 1} \text{Cos}(x - 1) = 1$.
Therefore

$$\lim\limits_{x \to 1} \frac{1 - \text{Cos}(x^2 - 1)}{1 - \text{Cos}^2(x - 1)} = \lim\limits_{x \to 1} \frac{(x + 1)^2}{1 + \text{Cos}(x - 1)} = \frac{4}{2} = 2.$$

EXERCISE (iii)
Evaluate

(a) $\lim\limits_{x \to 2} \dfrac{x^4 - 16}{x - 2}$,   (b) $\lim\limits_{x \to 2} \dfrac{x^2 - 3x + 4}{x^2 - 16}$,   (c) $\lim\limits_{x \to \infty} \dfrac{3x - 1}{x + 57}$,

(d) $\lim\limits_{x \to 0} \dfrac{\text{Sin } x^2}{\text{Sin } x}$,   (e) $\lim\limits_{x \to 2} \dfrac{\text{Sin } (x^2 - 4)}{x - 2}$,

(f) $\lim\limits_{x \to 0} \dfrac{\text{Cos} (\text{Sin } x)}{1 - x}$,   (g) $\lim\limits_{x \to \infty} x \, \text{Sin } \dfrac{1}{x}$.

3.3   Continuous functions

We have stressed before that a function $f(x)$ may have a limit as x tends to α without there being any connection between $\lim\limits_{x \to \alpha} f(x)$ and $f(\alpha)$.
Indeed, $f(x)$ may not even be defined for $x = \alpha$. Naturally it is of interest to investigate functions for which $\lim\limits_{x \to \alpha} f(x) = f(\alpha)$.

If $\lim\limits_{x \to \alpha} f(x)$ *exists, and is equal to* $f(\alpha)$, *then* $f(x)$ *is said to be*

continuous at $x = \alpha$.

Example 6   Show that x^3 is continuous for all values of x.

Solution   At $x = \alpha, f(\alpha) = \alpha^3$ and
$$|x^3 - \alpha^3| = |x - \alpha| \, |x^2 + \alpha x + \alpha^2| \leqslant |x - \alpha|(|x|^2 + |\alpha| \, |x| + |\alpha|^2).$$
If $|x - \alpha| < 1$, then $|x| < 1 + |\alpha|$ (and since $|\alpha| < 1 + |\alpha|$) this implies
$$|x^3 - \alpha^3| < |x - \alpha| \, . \, 3(1 + |\alpha|)^2.$$
Given $\epsilon > 0$ therefore, we have
$$|x^3 - \alpha^3| < \epsilon \quad \text{provided } |x - \alpha| < \delta = \min\left\{1, \frac{\epsilon}{3(1 + |\alpha|)^2}\right\}.$$

Thus x^3 is continuous at α. Since α was an arbitrarily chosen value of x, then x^3 is continuous everywhere.

EXERCISE (iv)
Show that x^n is continuous for all x if $n = 0, 1, 2, 3, \ldots$.

Example 7   Show that e^x is continuous for all values of x.

Solution   We must show that $e^x \to e^\alpha$ as $x \to \alpha$, or, put slightly differently, that $e^{x - \alpha} \to 1$ as $x \to \alpha$. In view of Proposition 3.3, this is equivalent to showing that $e^x \to 1$ as $x \to 0$. For this we can use the method of Example 2 (cf. Ex. (ii)(d)).

Immediately from Proposition 3.2 we obtain as a corollary

PROPOSITION 3.4
If $f(x)$ and $g(x)$ are continuous at $x = \alpha$, then so also are the functions $f(x) + g(x), f(x)g(x)$ and (if $g(\alpha) \neq 0$) $f(x)/g(x)$.

Note that, once we have shown that the function $f(x) = x$ is continuous for all x, then Proposition 3.4 tells us that all powers $x^n (n > 0)$ of x are continuous for all x, and hence also all polynomials $P(x)$ in x. This implies also continuity of all rational functions $P(x)/Q(x)$ for all x except those values α for which $Q(\alpha) = 0$.

EXERCISE (v)
Use the results of Ex. (ii) and the equalities of Chapter 2, Ex. (xxiii)(c) to show that Sin x and Cos x are continuous for all x.

3.4 Some properties of continuous functions

Throughout this section we shall suppose that $f(x)$ is continuous for all values of x lying within the range $a \leqslant x \leqslant b$ (the *closed* interval from a to b). The set of such x will be denoted by the symbol $[a, b]$, and the fact that "x belongs to the set $[a, b]$" will be indicated by the notation $x \in [a, b]$. Thus we consider $f(x)$ continuous for all $x \in [a, b]$. The square brackets here is standard notation—sometimes it is convenient to exclude the end points a and b, in which case the range $a < x < b$ (the *open* interval from a to b) will be denoted by (a, b).

The first property of our function is described by the Intermediate Value Theorem. Briefly, it states that all values between $f(a)$ and $f(b)$ are obtained as values of the function.

PROPOSITION 3.5   (Intermediate Value Theorem)
Suppose that $f(x)$ is continuous in $[a, b]$, and that $f(a) \neq f(b)$. Then every number between $f(a)$ and $f(b)$ is the value of $f(x)$ for some x in (a, b).

Before we prove Proposition 3.5 we shall consider a special case.

PROPOSITION 3.6

Suppose that $f(x)$ is continuous in $[a, b]$ and that $f(a)$ and $f(b)$ have opposite signs. Then $f(x) = 0$ for some x in (a, b).

Proof
For the sake of the proof we may assume, without any loss of generality, that $f(a) > 0$ and $f(b) < 0$. Let $h = b - a$. Then $h > 0$ and $a + h = b$. Let x_1 be the greatest of the numbers

$$a, \qquad a + \tfrac{1}{2}h, \qquad a + h$$

for which $f(x) > 0$. Then take x_2 to be the greatest among

$$a, \qquad a + \tfrac{1}{4}h, \qquad a + \tfrac{2}{4}h, \qquad a + \tfrac{3}{4}h, \qquad a + h$$

for which $f(x) > 0$. In general, take x_n to be the greatest among

$$a, \quad a + \frac{1}{2^n}h, \quad a + \frac{2}{2^n}h, \quad a + \frac{3}{2^n}h, \quad \ldots, \quad a + \frac{2^n - 1}{2^n}h, \quad a + h$$

for which $f(x) > 0$. Clearly then $a \leqslant x_1 \leqslant x_2 \leqslant \cdots \leqslant x_n \leqslant \cdots < b$, and so x_n converges to some α, say, where $\alpha \in [a, b]$. Since f is continuous at α, then

$$f(\alpha) = \lim_{n \to \infty} f(x_n), \quad \text{and} \quad f(\alpha) \geqslant 0 \quad \text{(Proposition 3.1 (iv))}.$$

But the sequence $x_n + (1/2^n)h$ also converges to α, while $f(x_n + (1/2^n)h) \leqslant 0$, and so

$$f(\alpha) = \lim_{n \to \infty} f\left(x_n + \frac{1}{2^n}h\right), \quad \text{and} \quad f(\alpha) \leqslant 0 \quad \text{(Proposition 3.1 (iv))}.$$

Therefore $f(\alpha) = 0$, and since $\alpha \neq a$, $\alpha \neq b$ (because $f(a) \neq 0$, $f(b) \neq 0$) then $\alpha \in (a, b)$.

Now the proof of Proposition 3.5 is easy.

Proof   (of Proposition 3.5)
Let ξ be any number between $f(a)$ and $f(b)$. Then $f(x) - \xi$ is continuous in $[a, b]$ and takes values of different signs at a and b. Therefore for some α in (a, b) we have $f(\alpha) - \xi = 0$, i.e., $f(\alpha) = \xi$.

Example 8   Show that there is a positive real solution of $x^7 = 2$.

Solution   The function $f(x) = x^7 - 2$ is continuous for all x, and in particular for $x \in [0, 2]$. Since $f(0) = -2$ and $f(2) = 126$, which are of different signs, then there is an α in $(0, 2)$ such that $f(\alpha) = 0$, i.e., such that $\alpha^7 = 2$.

EXERCISE (vi)
Show that, if $\alpha > 0$ and n is a positive integer, then $x^n = \alpha$ has a unique positive real solution. (For uniqueness, it may be useful to use the identity (1) of Chapter 1.)

The second property of our function assures us that the values it takes are within control.

PROPOSITION 3.7
If $f(x)$ is continuous in $[a, b]$, then it is bounded there.

Proof
As in the proof of Proposition 3.6, we consider the numbers

$$a, \quad a + \frac{1}{2^n} h, \quad a + \frac{2}{2^n} h, \quad a + \frac{3}{2^n} h, \quad \ldots, \quad a + \frac{2^n - 1}{2^n} h, \quad a + h,$$

and define x_n to be the greatest of these values for which $f(x)$ is bounded in $[a, x_n]$. Again we have $a \leqslant x_1 \leqslant x_2 \leqslant \cdots \leqslant x_n \leqslant \cdots \leqslant b$, and so x_n converges to some limit β, say. We remark that $x_n \leqslant \beta \leqslant b$. Since $f(x)$ is continuous at β, then (Proposition 3.1 (ii)) for some $\delta > 0$ $f(x)$ is bounded for $\beta - \delta < x < \beta + \delta$. But the interval $[\beta - \delta, \beta]$ must contain an x_n, and this implies that $f(x)$ is bounded in $[a, \beta + \delta]$. Suppose, if possible, that $\beta < b$. We can always find an integer N such that $1/2^N < \delta/h$, i.e., such that $h/2^N < \delta$, and so some number of the form $a + rh/2^N$ (where $1 \leqslant r \leqslant 2^N$) lies between β and $\beta + \delta$. Thus $f(x)$ is bounded in $[a, a + rh/2^N]$, and therefore $\beta < a + rh/2^N \leqslant x_N$. This is impossible since $x_N \leqslant \beta$. This contradiction leads us to conclude that $\beta = b$.

Note that it is vital for the validity of this proposition that $f(x)$ must be continuous at the end points a and b of the closed interval $[a, b]$. Continuity over the open interval (a, b) is not good enough, as the example $f(x) = 1/x$ shows. This is continuous for $0 < x < 1$, but is certainly not bounded over this range.

Now that we know that the set of values of $f(x)$ over $[a, b]$ is bounded, then (see comment about real numbers in §2.1), we know that these values have a least upper bound M, and a greatest lower bound m. The last property of continuous functions which we mention here states that these greatest and least values are achieved by the function. More explicitly,

PROPOSITION 3.8
Suppose $f(x)$ is continuous in $[a, b]$, and let M, m denote the least upper bound and greatest lower bound respectively of its set of values over $[a, b]$. Then there are numbers α, β in $[a, b]$ such that $f(\alpha) = M$, $f(\beta) = m$.

Proof
Given any $\epsilon > 0$, there exists an x_0 in $[a, b]$ such that $M - \epsilon < f(x_0)$ (by definition of least upper bound). Hence

$$\frac{1}{M - f(x_0)} > \frac{1}{\epsilon},$$

and so the function $1/(M - f(x))$ is not bounded in $[a, b]$, which implies it is not continuous there. But if $f(x) \neq M$ anywhere in $[a, b]$, then $1/(M - f(x))$ would be continuous there. Thus for some value of x, say $x = \alpha$ where $\alpha \in [a, b]$, we have $f(\alpha) = M$. The result for the greatest lower bound is proved similarly.

3.5 Derivative of a function

Once again we remind the reader that the aim of this chapter is put on a sound basis notions about functions which are to some extent already familiar. As mentioned in the introduction to the chapter, the reader is assumed to be more experienced than a complete novice, and our treatment is therefore in the nature of a recapitulation of existing knowledge.

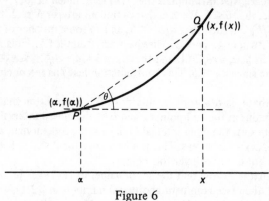

Figure 6

Consider a function $f(x)$ which we suppose is defined at and near the point $x = \alpha$. Then the function

$$\frac{f(x) - f(\alpha)}{x - \alpha}, \qquad x \neq \alpha, \tag{2}$$

measures the direction of the chord joining the points with cartesian coordinates $(\alpha, f(\alpha)), (x, f(x))$, for its value is $\tan \theta$. If this chord has a limit position as x tends to α then the limit will be called the *tangent* to the curve at P (Fig. 6). In this case the angle θ (and hence also, $\tan \theta$) will tend to a limit as $x \to \alpha$. These geometrical considerations prompt us to consider therefore the limit of the function (2) as $x \to \alpha$.

The limit $\lim\limits_{x \to \alpha} (f(x) - f(\alpha))/(x - \alpha)$ *(if it exists) is called the derivative of* $f(x)$ *at* $x = \alpha$, *and is denoted by* $f'(\alpha)$.

Another (equivalent) way of defining $f'(\alpha)$ of course is to take it to be

$$\lim_{h \to 0} \frac{f(\alpha + h) - f(\alpha)}{h}.$$

When $f(x)$ has a derivative at each point in some range of x then we can consider the function, denoted by $f'(x)$ or $df(x)/dx$, whose value at each point of the range is the derivative of $f(x)$ at that point. This $f'(x)$ is called the *derived function* (or briefly, *the derivative*) of $f(x)$.

The student will readily convince himself that not all functions (even continuous ones) possess a derivative. For example the function $f(x) = |x|$ behaves badly (in this respect) at $x = 0$. On the other hand, if $f(x)$ has a derivative at $x = \alpha$, then $f(x)$ is continuous there, for

$$\lim_{x \to \alpha} \{f(x) - f(\alpha)\} = \lim_{x \to \alpha} \left\{ \frac{f(x) - f(\alpha)}{x - \alpha} \right\} (x - \alpha)$$

$$= f'(\alpha) \lim_{x \to \alpha} (x - \alpha) = 0.$$

EXERCISE (vii)
Show that the following functions are continuous, but do not have a derivative at $x = 0$:

(a) $|x|$;   (b) $|x|/(1 + x^2)$;

(c) $f(x) = x$ for $x < 0$, $f(x) = x^2$ for $x \geqslant 0$.

Example 9   Find the derivative of $f(x) = x^n$, where n is a positive integer.

Solution   We consider the derivative at a point $x = \alpha$. First note that, for $x \neq \alpha$

$$\frac{x^n - \alpha^n}{x - \alpha} = x^{n-1} + x^{n-2}\alpha + x^{n-3}\alpha^2 + \cdots + x\alpha^{n-2} + \alpha^{n-1}.$$

Since each of the terms in the right hand sum has limit α^{n-1} as $x \to \alpha$, then $\lim_{x \to \alpha} (x^n - \alpha^n)/(x - \alpha) = n\alpha^{n-1}$, i.e., $f'(x) = nx^{n-1}$ everywhere.

Example 10   Show that the derivative of e^x is e^x.

Solution   Our solution uses the result of Ex. (ii)(e), rather than the term by term differentiation of §2.8. At any point x, the derivative is

$$\lim_{h \to 0} \frac{e^{x+h} - e^x}{h} = e^x \left(\lim_{h \to 0} \frac{e^h - 1}{h} \right) = e^x$$

by the exercise quoted.

EXERCISE (viii)
(a) Find the derivative of $f(x) = x^{-n}$ (n is a positive integer) at α where $\alpha \neq 0$.
(b) Use the identities of Chapter 2, Ex. (xxiii)(c) to show that $\text{Sin } x$ has derivative $\text{Cos } x$, and $\text{Cos } x$ has derivative $-\text{Sin } x$.

3.6 Useful rules for differentiation

In practice functions we encounter are more complicated than those of Examples 9 and 10. We give here rules for obtaining the derivative of composite functions.

PROPOSITION 3.9
If $f(x)$ and $g(x)$ have derivatives for some range of x, and if c is a constant, then

  (i) *$cf(x)$ has derivative $cf'(x)$,*
 (ii) *$f(x) + g(x)$ has derivative $f'(x) + g'(x)$,*
(iii) *$f(x)g(x)$ has derivative $f'(x)g(x) + f(x)g'(x)$,*
(iv) *$f(x)/g(x)$ has derivative $\{f'(x)g(x) - f(x)g'(x)\}/\{g(x)\}^2$*
provided $g(x) \neq 0$.

Proof
(i) is a consequence of (iii) and the fact that the derivative of a constant function is 0. (ii) is an easy consequence of Proposition 3.2 (ii).
   For (iii)

$$\lim_{h \to 0} \frac{f(x+h)g(x+h) - f(x)g(x)}{h}$$

$$= \lim_{h \to 0} \left\{ g(x+h)\frac{f(x+h) - f(x)}{h} + f(x)\frac{g(x+h) - g(x)}{h} \right\}$$

$$= \lim_{h \to 0} g(x+h) \cdot \lim_{h \to 0} \frac{f(x+h) - f(x)}{h}$$

$$+ f(x) \cdot \lim_{h \to 0} \frac{g(x+h) - g(x)}{h}$$

$$= g(x)f'(x) + f(x)g'(x).$$

Finally, if $g(x) \neq 0$,

$$\lim_{h \to 0} \frac{1}{h} \left\{ \frac{f(x+h)}{g(x+h)} - \frac{f(x)}{g(x)} \right\}$$

$$= \lim_{h \to 0} \left\{ \frac{g(x) \dfrac{f(x+h) - f(x)}{h} - f(x) \dfrac{g(x+h) - g(x)}{h}}{g(x+h)g(x)} \right\} \Bigg/$$

$$= \frac{g(x) \cdot \lim_{h \to 0} \dfrac{f(x+h) - f(x)}{h} - f(x) \cdot \lim_{h \to 0} \dfrac{g(x+h) - g(x)}{h}}{g(x) \cdot \lim_{h \to 0} g(x+h)}$$

$$= \{g(x)f'(x) - f(x)g'(x)\}/\{g(x)\}^2.$$

Example 11   Find the derivative of $\dfrac{x \, \mathrm{Sin}\, x - \mathrm{Cos}\, x}{x^2}$.

Solution   The numerator has derivative $x \, \mathrm{Cos}\, x + \mathrm{Sin}\, x + \mathrm{Sin}\, x$; the denominator has derivative $2x$. Hence, by Proposition 3.9 (iv) our function has the derivative $\{x^2(x \, \mathrm{Cos}\, x + 2 \, \mathrm{Sin}\, x) - (x \, \mathrm{Sin}\, x - \mathrm{Cos}\, x)2x\}/x^4$, which on simplifying gives $\{(x^2 + 2) \, \mathrm{Cos}\, x\}/x^3$.

The next rule deals with functions of the form $f(g(x))$. (Consider, for example, the functions $f(t) = \mathrm{Sin}\, t, g(x) = x^3 + 2x$. Then $f(g(x)) = \mathrm{Sin}\,(x^3 + 2x)$.)

PROPOSITION 3.10   **(The Chain Rule)**
Suppose that $g(x)$ is differentiable at $x = \alpha$, and $f(t)$ is differentiable at $t = g(\alpha)$. Then $f(g(x))$, regarded as a function of x, is differentiable at $x = \alpha$ with derivative $f'(g(\alpha))g'(\alpha)$ there.

Proof
The usual "proof" given for this important theorem is reproduced here. The student is warned that it contains a mistake, which is explained below. The fallacious proof goes:
We wish to find

$$\lim_{h \to 0} \frac{f(g(\alpha + h)) - f(g(\alpha))}{h}.$$

First put $k = g(\alpha + h) - g(\alpha)$, and note that since $g(x)$ is continuous at $x = \alpha$, then $k \to 0$ as $h \to 0$. Thus

$$\lim_{h \to 0} \frac{f(g(\alpha + h)) - f(g(\alpha))}{h}$$

$$= \lim_{h \to 0} \left(\frac{f(g(\alpha) + k) - f(g(\alpha))}{k} \cdot \frac{g(\alpha + h) - g(\alpha)}{h} \right)$$

$$= \lim_{h \to 0} \frac{f(g(\alpha) + k) - f(g(\alpha))}{k} \cdot \lim_{h \to 0} \frac{g(\alpha + h) - g(\alpha)}{h}$$

$$= \lim_{k \to 0} \frac{f(g(\alpha) + k) - f(g(\alpha))}{k} \cdot \lim_{h \to 0} \frac{g(\alpha + h) - g(\alpha)}{h}$$

$$= f'(g(\alpha))g'(\alpha).$$

As mentioned above, this proof is fallacious—it ignores the possibility that k may be zero (for example when $g(x)$ is a constant function). The correct proof follows.

From the definition of the derivative we have

$$g(\alpha + h) - g(\alpha) = \{g'(\alpha) + \epsilon(h)\} h$$

where $\epsilon(h) \to 0$ as $h \to 0$. If we put $k = g(\alpha + h) - g(\alpha)$, then we see that $k \to 0$ as $h \to 0$. Again from the definition,

$$f(g(\alpha) + k) - f(g(\alpha)) = \{f'(g(\alpha)) + \eta(k)\}k$$

where $\eta(k) \to 0$ as $k \to 0$, and hence as $h \to 0$. Finally,

$$\frac{f(g(\alpha + h)) - f(g(\alpha))}{h} = \{f'(g(\alpha)) + \eta(k)\}\{g'(\alpha) + \epsilon(h)\}.$$

and the right-hand side of this equation tends to $f'(g(\alpha))g'(\alpha)$ as $h \to 0$.

The result of this proposition is often stated in the form

$$\frac{df}{dx} = \frac{df}{dt}\frac{dt}{dx}.$$

Example 12   Find the derivative of $\text{Sin}(x^3 + 2x)$.

Solution   Sin t has derivative Cos t, and $x^3 + 2x$ has derivative $3x^2 + 2$. Therefore $\text{Sin}(x^3 + 2x)$ has derivative $(3x^2 + 2)\text{Cos}(x^3 + 2x)$.

Example 13   Find the derivative of $(7x^2 + 3x)^4$.

Solution   Put $f(t) = t^4$ and $g(x) = 7x^2 + 3x$. Then $f(g(x)) = (7x^2 + 3x)^4$. Since $f'(t) = 4t^3$ and $g'(x) = 14x + 3$, then the required derivative is $4(7x^2 + 3x)^3(14x + 3)$.

EXERCISE (ix)
Show that the functions

(a) e^{-x},   (b) $\sinh x$,   (c) $\cosh x$,   (d) $e^{\mathrm{Sin}\, x}$,   (e) e^{x^2+1},
(f) $\mathrm{Sin}\,(\mathrm{Sin}\, x)$,   (g) $e^{x\,\mathrm{Cos}\, x}$,

have derivatives

(a') $-e^{-x}$,   (b') $\cosh x$,   (c') $\sinh x$,   (d') $\mathrm{Cos}\, x \cdot e^{\mathrm{Sin}\, x}$,
(e') $2xe^{x^2+1}$,   (f') $\mathrm{Cos}\, x \cdot \mathrm{Cos}\,(\mathrm{Sin}\, x)$,
(g') $(\mathrm{Cos}\, x - x\,\mathrm{Sin}\, x)e^{x\,\mathrm{Cos}\, x}$ respectively.

EXERCISE (x)
Find the derivatives of the functions

(a) $e^{\mathrm{Sin}\, x^2}$,   (b) $\mathrm{Cos}\,(e^x)$,   (c) $(3x+2)^3$,   (d) $e^{(3x+2)^3}$,
(e) $\mathrm{Sinh}\,(e^{(3x+2)^3})$.

In §2.8 we defined the function $\log x$, for $x > 0$, by taking $\log x$ to be that unique number for which $e^{\log x} = x$ In general, if a function $f(x)$ is defined for $a < x < b$ and has the property that $x_1 \neq x_2$ implies $f(x_1) \neq f(x_2)$, then we can define a new function (denoted by $f^{-1}(y)$) which takes the value x at the point $f(x)$, i.e., $f^{-1}(f(x)) = x$. (Note that we must not confuse $f^{-1}(y)$ with the function $1/f(x)$.) For example, the function $\sin x$ never takes the same value twice in the range. $-\pi/2 < x < \pi/2$. Thus $\sin^{-1} y$ is defined for $-1 < y < 1$, and since $\sin \pi/4 = 1/\sqrt{2}$, then $\sin^{-1}(1/\sqrt{2}) = \pi/4$.

The function $f^{-1}(y)$ is called the *inverse function* of $f(x)$. Graphically, we may consider it thus. The graph of the function $f(x)$ is drawn as usual with the x-axis representing values of the variable and the y-axis representing the values of the function. Then the same graph will depict $f^{-1}(y)$ provided the roles of the axes are reversed: the y-axis now represents the values of the variable and the x-axis represents the values of the function $f^{-1}(y)$.

PROPOSITION 3.11   **(The Inverse Function Theorem)**
Suppose that $y = f(x)$ is strictly increasing and continuous in $a < x < b$. Then the inverse function $x = f^{-1}(y)$ is defined in $f(a) < y < f(b)$, and is strictly increasing and continuous.

Proof
It is easy to show that $f^{-1}(y)$ is strictly increasing. We use it to prove continuity for $f^{-1}(y)$. Take y_0 satisfying $f(a) < y_0 < f(b)$, and put $x_0 = f^{-1}(y_0)$. Then $a < x_0 < b$. Given $\epsilon > 0$, choose x_1, x_2 in (a, b) so that $x_0 - \epsilon < x_1 < x_0 < x_2 < x_0 + \epsilon$. Thus $f(a) < f(x_1) < y_0 < f(x_2) < f(b)$. If we take $\delta = \min\{y_0 - f(x_1), f(x_2) - y_0\}$ then $|y - y_0| < \delta$ implies $f(x_1) < y < f(x_2)$ which gives $x_1 < f^{-1}(y) < x_2$, and hence $|f^{-1}(y) - x_0| < \epsilon$.

PROPOSITION 3.12

Suppose that $f(x)$ is strictly increasing and differentiable for $a < x < b$, and that $f'(x) \neq 0$ in this range. Then the inverse function $f^{-1}(y)$ is differentiable in $f(a) < y < f(b)$ with derivative $1/f'(f^{-1}(y))$. Since $x = f^{-1}(y)$ if $y = f(x)$, then this proposition can be briefly stated in the form

$$\frac{d}{dy} f^{-1}(y) = \frac{1}{f'(x)}$$

or

$$\frac{dx}{dy} = 1 \bigg/ \frac{dy}{dx}.$$

Proof
For x in the range $a < x < b$, put $y = f(x)$, and for $h \neq 0$ put $y + k = f(x + h)$. Then $k \neq 0$, while $f^{-1}(y) = x$ and $f^{-1}(y + k) = x + h$. This gives

$$\frac{f^{-1}(y + k) - f^{-1}(y)}{k} = \frac{h}{k}.$$

Now

$$\frac{f(x + h) - f(x)}{h} = \frac{k}{h} \to f'(x)(\neq 0) \quad \text{as } h \to 0,$$

and so $h/k \to 1/f'(x)$ as $h \to 0$. But from the continuity of $f^{-1}(y)$, we have $h \to 0$ as $k \to 0$. Therefore $h/k \to 1/f'(x)$ as $k \to 0$.

Example 14   Find the derivative of $\log x$, for $x > 0$.

Solution   Put $y = \log x$. We wish to find dy/dx. Since $e^y = x$, then by finding the derivatives of both sides of the equality we have (using the chain rule) $e^y(dy/dx) = 1$. Therefore

$$\frac{dy}{dx} = \frac{1}{e^y} = \frac{1}{x}.$$

Example 15   Find the derivatives of $\sin^{-1} x$, for $-1 < x < 1$.

Solution   Put $y = \sin^{-1} x$. Then $\sin y = x$, and by the chain rule $\cos y(dy/dx) = 1$. This gives

$$\frac{dy}{dx} = \frac{1}{\cos y} = \frac{1}{\sqrt{(1 - x^2)}}.$$

EXERCISE (xi)
Show that the functions

(a) $\cos^{-1} x$,　　(b) $\tan^{-1} x$,　　(c) $\cosh^{-1} x$,　　(d) $\sinh^{-1} x$,

have derivatives

(a′) $-1/\sqrt{(1-x^2)}$,　　(b′) $1/(1+x^2)$,　　(c′) $1/\sqrt{(x^2-1)}$,
(d′) $1/\sqrt{(1+x^2)}$ respectively.

Note that the inverse functions of the circular functions $\sin x$, $\cos x$, $\tan x$ are sometimes denoted by arc sin x, arc cos x, arc tan x respectively.

EXERCISE (xii)
Find the derivatives of the functions

(a) $\cos^{-1}(\sin x)$,　　(b) $\sin^{-1} x^2$,　　(c) $\log(\sin^{-1} x)$,

stating for which ranges of x these functions are defined.

So far we have concentrated on the process of obtaining the derivative $f'(x)$ of a function $f(x)$. It is quite possible that $f'(x)$ may itself be differentiable, and *its* derivative will then be denoted by $f''(x)$ or sometimes, $f^{(2)}(x)$. If the function $f(x)$ can be differentiated n times we say it has a derivative of *order n* which we denote by $f^{(n)}(x)$. From Proposition 3.9 (i), (ii), if $f(x)$ and $g(x)$ have derivatives of order n then

$$\{cf\}^{(n)}(x) = cf^{(n)}(x),$$
$$\{f+g\}^{(n)}(x) = f^{(n)}(x) + g^{(n)}(x).$$

For the nth derivative of the product $f(x)g(x)$ we have

PROPOSITION 3.13 **(Leibniz's rule)**
Suppose that $f(x)$ and $g(x)$ have derivatives of order n in some common range. Then $h(x) = f(x)g(x)$ also has a derivative of order n in the range, and

$$h^{(n)}(x) = \sum_{r=0}^{n} \binom{n}{r} f^{(r)}(x) g^{(n-r)}(x). \tag{3}$$

Proof
The proof is by induction on n. The case $n = 1$ is given by Proposition 3.9 (iii). Suppose (induction hypothesis) that the formula (3) holds.

Then differentating term by term, we obtain

$$h^{(n+1)} = \sum_{r=0}^{n} \binom{n}{r} \{f^{(r)}g^{(n+1-r)} + f^{(r+1)}g^{(n-r)}\}$$

$$= \sum_{r=0}^{n} \binom{n}{r} f^{(r)}g^{(n+1-r)} + \sum_{r=0}^{n} \binom{n}{r} f^{(r+1)}g^{(n-r)}$$

$$= f \cdot g^{(n+1)} + \sum_{r=1}^{n} \binom{n}{r} f^{(r)}g^{(n+1-r)}$$

$$+ \sum_{r=1}^{n} \binom{n}{r-1} f^{(r)}g^{(n+1-r)} + f^{(n+1)} \cdot g$$

$$= f \cdot g^{(n)} + \sum_{r=1}^{n} \left\{\binom{n}{r} + \binom{n}{r-1}\right\} f^{(r)} \cdot g^{(n+1-r)} + f^{(n+1)} \cdot g$$

$$= f \cdot g^{(n+1)} + \sum_{r=1}^{n} \binom{n+1}{r} f^{(r)}g^{(n+1-r)} + f^{(n+1)}g,$$

Since $\binom{n}{r} + \binom{n}{r-1} = \binom{n+1}{r}$. This completes the proof, as we have

proved (3) with n replaced by $n + 1$.

Example 16   Find the nth derivative of $x^3 \sin x$.

Solution   If we take $f(x) = x^3$ and $g(x) = \sin x$, then in obtaining the nth derivative of $f(x)g(x)$ by Leibniz's rule we notice that we need go no further than the fourth term, since $f^{(r)}(x) = 0$, for $r \geqslant 4$. Accordingly, we see that the nth derivative is

$$x^3 (-1)^m \sin x + \binom{2m}{1} 3x^2 (-1)^{m-1} \cos x + \binom{2m}{2} 6x(-1)^{m-1}\sin x$$

$$+ \binom{2m}{3} 6(-1)^{m-2} \cos x \qquad \text{if } n = 2m,$$

and is

$$x^3(-1)^m \cos x + \binom{2m+1}{1} 3x^2(-1)^m \sin x + \binom{2m+1}{2} 6x(-1)^{m-1} \cos x$$

$$+ \binom{2m+1}{3} 6(-1)^{m-1} \sin x \quad \text{if } n = 2m+1.$$

EXERCISE (xiii)

(a) Find the nth derivative of $x^2 \sin (ax + b)$.

(b) Find the fourth derivative of $x^2 \log x$.

(c) Show that the fifth derivative of $x^2 e^x$ is $e^x(x^2 + 10x + 20)$.

(d) Find the nth derivative of $x^3/(x^2 - x - 2)$ (Hint: use partial fractions).

CHAPTER 4

Mean Value and Taylor's Theorems

4.1 First Mean Value Theorem

It is easy to prove from the definition that, if $f(x)$ is constant through-out an interval $a < x < b$, then $f'(x) = 0$ there. That the converse is also true is intuitively obvious—though the proof is not an entirely trivial matter (Example 1).

Another intuitively obvious observation is that, if the graph of a function $f(x)$ has a tangent at all points in the open interval (a, b), and if $f(a) = f(b)$, then somewhere in (a, b) the tangent is horizontal. This fundamental and useful theorem is

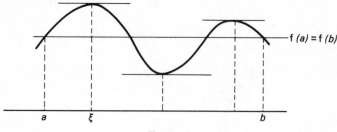

Figure 7

known as *Rolle's theorem.*

PROPOSITION 4.1　**(Rolle's Theorem)**
Suppose that $f(x)$ is a function which is continuous in $a \leqslant x \leqslant b$ and differentiable in $a < x < b$. Suppose also that $f(a) = f(b)$. Then there is a number ξ, with $a < \xi < b$, such that $f'(\xi) = 0$.

Proof
We recall (Propositions 3.7, 3.8) that $f(x)$ is bounded in $[a, b]$ and attains its bounds M, m there. Quite clearly, if $M = m$, then $f(x)$ is constant throughout $[a, b]$ and (as observed above) $f'(x) = 0$ for every x in (a, b). Suppose then that $M \neq m$. This implies that $f(a) = f(b) < M$ or $m < f(a) = f(b)$. In the first case we see that (Proposition 3.8) there exists ξ in (a, b) such that $f(a) = f(b) < f(\xi) = M$. Now by hypothesis

64

$f'(\xi)$ exists. But $(f(\xi + h) - f(\xi))/h \geqslant 0$ if $h < 0$ and $(f(\xi + h) - f(\xi))/h < 0$ if $h \geqslant 0$. Therefore its limit (as $h \to 0$) must be 0, i.e. $f'(\xi) = 0$. The other case when $m < f(a) = f(b)$ is proved similarly.

An important generalization of Rolle's theorem, for the case when $f(a) \neq f(b)$, is

PROPOSITION 4.2   **(First Mean Value Theorem)**
Suppose that $f(x)$ is continuous in $a \leqslant x \leqslant b$ and differentiable in $a < x < b$. Then there is a number ξ, satisfying $a < \xi < b$, such that

$$\frac{f(b) - f(a)}{b - a} = f'(\xi). \tag{1}$$

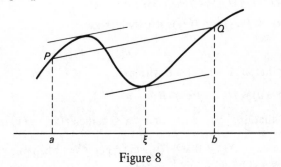

Figure 8

Proof
Let $F(x)$ be the function defined by

$$F(x) = f(b) - f(x) - (b - x)\lambda,$$

where λ is a constant chosen to make $F(a) = 0$, i.e. $\lambda = (f(b) - f(a))/(b - a)$. Then

$$F'(x) = -f'(x) + \lambda.$$

Since $F(a) = 0 = F(b)$, then the conditions of Rolle's theorem are satisfied, and so there exists ξ, with $a < \xi < b$, such that $F'(\xi) = 0$. In other words

$$f'(\xi) = \lambda = \frac{f(b) - f(a)}{b - a}.$$

To interpret this graphically we refer to Fig. 8. The First Mean Value Theorem states that, between a and b there is at least one point ξ at which the tangent to the curve is parallel to the chord PQ.

Example 1   Show that, if $f'(x) = 0$ throughout $a < x < b$, then $f(x)$ is constant there.

Solution   Take x_1, x_2 such that $a < x_1 < x_2 < b$. By the First Mean Value Theorem (MVT) applied to the interval $x_1 \leqslant x \leqslant x_2$,

$$f(x_2) - f(x_1) = (x_2 - x_1)f'(\xi) = 0, \text{ since } f'(\xi) = 0.$$

Hence $f(x_1) = f(x_2)$ for all points x_1, x_2 in the open interval (a, b).

Example 2   Show that if $f'(x) > 0$ throughout $a < x < b$, then $f(x)$ is strictly increasing for $a \leqslant x \leqslant b$.

Solution   Suppose $a \leqslant x_1 < x_2 \leqslant b$. Then by MVT we have

$$f(x_2) - f(x_1) = (x_2 - x_1)f'(\xi) > 0.$$

Hence $f(x_1) < f(x_2)$ and $f(x)$ is strictly increasing.

EXERCISE (i)
(a) Show that, if $A \leqslant f'(x) \leqslant B$ whenever $a < x < b$, then

$$A(b - a) \leqslant f(b) - f(a) \leqslant B(b - a).$$

(b) Show that, if $f'(x) = g'(x)$ for $a < x < b$, then $f(x) - g(x)$ is a constant for $a \leqslant x \leqslant b$.
(c) Show that, if $f(x) \neq 0$ and $f(x)g'(x) = g(x)f'(x)$ whenever $a < x < b$, then $g(x)$ is a constant multiple of $f(x)$.

Example 3
Show that, if $a < b$ then

$$\frac{b - a}{1 + b^2} < \tan^{-1} b - \tan^{-1} a < \frac{b - a}{1 + a^2}.$$

Deduce that

$$\frac{\pi}{4} + \frac{10}{221} < \tan^{-1} \frac{11}{10} < \frac{\pi}{4} + \frac{10}{200}.$$

Solution   Put $f(x) = \tan^{-1} x$. Then $f'(x) = 1/(1 + x^2)$ and by MVT, there is a ξ satisfying $a < \xi < b$ such that

$$\tan^{-1} b - \tan^{-1} a = \frac{b - a}{1 + \xi^2}.$$

Since $a < \xi < b$ then

$$\frac{b - a}{1 + b^2} < \frac{b - a}{1 + \xi^2} < \frac{b - a}{1 + a^2},$$

which yields the desired result. Now take $b = 11/10$, $a = 1$. Then $\tan^{-1} 1 = \pi/4$ and

$$\frac{b-a}{1+b^2} = \frac{10}{221}, \quad \frac{b-a}{1+a^2} = \frac{10}{200}.$$

Therefore

$$\frac{10}{221} < \tan^{-1} \frac{11}{10} - \frac{\pi}{4} < \frac{10}{200},$$

and hence

$$\frac{\pi}{4} + \frac{10}{221} < \tan^{-1} \frac{11}{10} < \frac{\pi}{4} + \frac{10}{200}.$$

EXERCISE (ii)

(a) Show that $\pi/6 + \sqrt{3}/15 < \sin^{-1} 3/5 < \pi/6 + 1/8$.

(b) If $0 < a < b$, show that $(b-a)/b < \log b/a < (b-a)/a$. Deduce that $1/6 < \log 1.2 < 1/5$.

Another generalization, this time of the MVT, is known as *Cauchy's form of the Mean Value Theorem.* This is

PROPOSITION 4.3   **(Cauchy's MVT)**
Suppose $f(x)$ and $g(x)$ are continuous for $a \leqslant x \leqslant b$ and differentiable for $a < x < b$. Suppose also that $g'(x) \neq 0$ for $a < x < b$. Then there is a number ξ, satisfying $a < \xi < b$, such that

$$\frac{f(b) - f(a)}{g(b) - g(a)} = \frac{f'(\xi)}{g'(\xi)}.$$

Proof
We remark that, since $g'(x) \neq 0$, then by Rolle's theorem $g(a) \neq g(b)$. Define $F(x)$ by

$$F(x) = f(b) - f(x) - \{g(b) - g(x)\}\lambda$$

where λ is a constant chosen to make $F(a) = 0$, i.e.,

$$\lambda = \frac{f(b) - f(a)}{g(b) - g(a)}.$$

Then

$$F'(x) = -f'(x) + g'(x)\lambda.$$

Since $F(a) = 0 = F(b)$, then by Rolle's theorem there is a ξ between a and b such that $F'(\xi) = 0$. In other words

$$f'(\xi) = g'(\xi)\lambda = g'(\xi)\left(\frac{f(b) - f(a)}{g(b) - g(a)}\right),$$

which yields the desired result.

It is worth comparing this result and its proof with that of the First Mean Value Theorem. Just take $g(x)$ to be the function $x - a$.

EXERCISE (iii)

Suppose that $f(x)$ and $g(x)$ are n times differentiable for $a \leqslant x \leqslant b$, $g'(x), \ldots, g^{(n)}(x)$ are not zero for $a < x < b$, $g'(a) = \cdots = g^{(n-1)}(a) = 0$, $f'(a) = \cdots = f^{(n-1)}(a) = 0$ and $g^{(n)}(a) \neq 0$. Show then that, for some ξ satisfying $a < \xi < b$,

$$\frac{f(b) - f(a)}{g(b) - g(a)} = \frac{f^{(n)}(\xi)}{g^{(n)}(\xi)} .$$

Reformulate this result in the case when $g(x) = (x - a)^n$.

4.2   Taylor's theorem

As we have seen from Example 3, one of the uses of the First Mean Value Theorem is to give approximate values of functions. Naturally it would be very desirable to be able to extend this process so that the resulting approximation is a good one. This is achieved by *Taylor's theorem*. Before stating this theorem (which is also known as the *Nth Mean Value Theorem*) we first consider the case $N = 2$.

PROPOSITION 4.4   (Second Mean Value Theorem)

Suppose that $f(x)$ is a function such that $f'(x)$ exists and is continuous in $a \leqslant x \leqslant b$ and differentiable in $a < x < b$. Then there is a number ξ, satisfying $a < \xi < b$, such that

$$f(b) = f(a) + \frac{(b - a)}{1!} f'(a) + \frac{(b - a)^2}{2!} f''(\xi). \tag{2}$$

Proof

Let $F(x)$ be the function defined by

$$F(x) = f(b) - f(x) - \frac{(b - x)}{1!} f'(x) - \frac{(b - x)^2}{2!} \lambda$$

where λ is a constant chosen to make $F(a) = 0$, i.e.

$$\lambda = \frac{2!}{(b - a)^2} \left\{ f(b) - f(a) - \frac{(b - a)}{1!} f'(a) \right\}. \tag{3}$$

Then

$$F'(x) = (b - x)(-f''(x) + \lambda).$$

Since $F(a) = 0 = F(b)$, then by Rolle's theorem there is a ξ with $a < \xi < b$ such that

$$F'(\xi) = (b - \xi)(-f''(\xi) + \lambda) = 0,$$

i.e. such that

$$\lambda = f''(\xi).$$

Note that the proof follows the same lines as that of the First Mean Value Theorem. In fact if we write the equation (1) in the form

$$f(b) = f(a) + (b - a)f'(\xi) \tag{4}$$

then the equations (4) and (2) indicate that there is a general pattern for results of this type. This extension to the general case is *Taylor's theorem.*

PROPOSITION 4.5   **(Taylor's Theorem)**
Suppose that $f(x)$ is a function such that $f^{(N-1)}(x)$ exists and is continuous in $a \leqslant x \leqslant b$ and differentiable in $a < x < b$. Then there is a number ξ, satisfying $a < \xi < b$, such that

$$f(b) = f(a) + \frac{(b-a)}{1!}f'(a) + \frac{(b-a)^2}{2!}f''(a) + \cdots$$

$$+ \frac{(b-a)^{N-1}}{(N-1)!}f^{(N-1)}(a) + \frac{(b-a)^N}{N!}f^{(N)}(\xi). \tag{5}$$

Proof
Let $F(x)$ be the function defined by

$$F(x) = f(b) - f(x) - \frac{(b-x)}{1!}f'(x) \quad - \cdots$$

$$- \frac{(b-x)^{N-1}}{(N-1)!}f^{(N-1)}(x) - \frac{(b-x)^N}{N!}\lambda$$

where λ is a constant chosen to make $F(a) = 0$. In other words,

$$\lambda = \frac{N!}{(b-a)^N}\left\{f(b) - f(a) - \frac{(b-a)}{1!}f'(a) - \cdots\right.$$

$$\left. - \frac{(b-a)^{N-1}}{(N-1)!}f^{(N-1)}(a)\right\}. \tag{6}$$

Then

$$F'(x) = \frac{(b-x)^{N-1}}{(N-1)!}(-f^{(N)}(x) + \lambda).$$

Since $F(a) = 0 = F(b)$, then by Rolle's theorem there is a ξ strictly between a and b such that

$$F'(\xi) = \frac{(b-\xi)^{N-1}}{(N-1)!}(-f^{(N)}(\xi) + \lambda) = 0,$$

i.e. such that

$$\lambda = f^{(N)}(\xi).$$

This, with equation (6), gives the result.

If we prefer, we may of course put $h = b - a$, and rewrite the equation (5) in the form

$$f(a + h) = f(a) + \frac{h}{1!} f'(a) + \frac{h^2}{2!} f''(a) + \cdots + \frac{h^{N-1}}{(N-1)!} f^{(N-1)}(a)$$

$$+ \frac{h^N}{N!} f^{(N)}(a + \theta h) \tag{7}$$

where $0 < \theta < 1$. At first sight it would seem that the validity of (7) was asserted by Taylor's theorem only for $h > 0$. In fact closer examination of the proof shows that nowhere do we assume that $b - a > 0$. Hence Taylor's theorem holds provided $f^{(N-1)}(x)$ is continuous "between a and b" and is differentiable "strictly between a and b", and so (5) is valid even if $b < a$. This means that (7) is valid irrespective of the sign of h.

A special case of Taylor's theorem (known as *Maclaurin's Theorem*) is obtained by taking $a = 0$. In the form (7), and writing x instead of h, this is

$$f(x) = f(0) + \frac{f'(0)}{1!} x + \frac{f''(0)}{2!} x^2 + \cdots + \frac{f^{(N-1)}(0)}{(N-1)!} x^{N-1}$$

$$+ \frac{f^{(N)}(\theta x)}{N!} x^N, \tag{8}$$

where $0 < \theta < 1$.

The equation (5) (or equivalently, (7)) is known as the *Taylor expansion* of $f(x)$ about $x = a$. The term $((b-a)^N/N!)f^{(N)}(\xi)$ (or $(h^N/N!)f^{(N)}(a + \theta h)$) is called the *remainder after N terms*, and is denoted by R_N. If we put

$$S_N = f(a) + \frac{h}{1!} f'(a) + \frac{h^2}{2!} f''(a) + \cdots + \frac{h^{N-1}}{(N-1)!} f^{(N-1)}(a) \tag{9}$$

then S_N is a polynomial in h whose coefficients depend only on $f(x)$ and its derivatives at a. Furthermore we have

$$f(a + h) = S_N + R_N. \tag{10}$$

If we know that R_N is insignificantly small (perhaps because h is small, or N is large) then we see from (10) that we may take S_N to be an approximation to $f(x)$, valid near $x = a$. The only problem is to decide whether R_N is small.

Example 4   Use the Third MVT to estimate $\tan^{-1} 11/10 - \pi/4$.

Solution   Put $f(x) = \tan^{-1} x$, then $f'(x) = 1/(1 + x^2)$, $f''(x) = -2x/(1 + x^2)^2$, and $f'''(x) = (6x^2 - 2)/(1 + x^2)^3$. Using Taylor's expansion about $x = 1$ with $N = 3$, we have

$$\tan^{-1} \frac{11}{10} = \frac{\pi}{4} + \frac{1}{10} \frac{1}{2} + \frac{1}{2!} \left(\frac{1}{10}\right)^2 \left(\frac{-2}{4}\right) + \frac{1}{3!} \left(\frac{1}{10}\right)^3 \left(\frac{6\xi^2 - 2}{(1 + \xi^2)^3}\right),$$

where $1 < \xi < 1.1$. Hence

$$\tan^{-1} \frac{11}{10} - \frac{\pi}{4} = 0.0475 + R_3,$$

where

$$\frac{1}{6000} \cdot \frac{4}{11} < \frac{1}{6000} \frac{6 - 2}{(1 + 1.1^2)^3} < R_3 = \frac{1}{3!} \left(\frac{1}{10}\right)^3 \left(\frac{6\xi^2 - 2}{(1 + \xi^2)^3}\right)$$

$$< \frac{1}{6000} \frac{6(1.1)^2 - 2}{(1 + 1)^3} < \frac{1}{6000} \cdot \frac{5.26}{8},$$

i.e. $0.0000606 < R_3 < 0.0001096$. Therefore

$$0.0475606 < \tan^{-1} \frac{11}{10} - \frac{\pi}{4} < 0.0476096.$$

EXERCISE (iv)
Use the Taylor expansion (with $N = 3$) of $(1 + x)^{1/5}$ about $x = 0$ to show that, to four decimal places, $(1.1)^{1/5} = 1.0192$.

EXERCISE (v)
Write out Maclaurin's expansion for the following functions

(a) $\sin x$,   (b) $\cos x$,   (c) $\log(1 + x)$,   (d) $\sinh x$,

(e) $\cosh x$, giving in each case an explicit formula for R_N.

EXERCISE (vi)
Compute to four decimal places the values of

(a) $(0.91)^{1/3}$,   (b) $\log(0.9)$.

EXERCISE (vii)
Approximate to $\log(1 + x)$ by a polynomial of third degree and show that, for $0 < x < \frac{1}{2}$, the error involved is less than 2^{-6}.

Example 5   Show that e is irrational.

Solution   Let $f(x) = e^x$. Since $f^{(n)}(x) = e^x$, then by Maclaurin's theorem and noting that $f^{(n)}(0) = 1$, we have

$$e^x = 1 + \frac{x}{1!} + \frac{x^2}{2!} + \cdots + \frac{x^{N-1}}{(N-1)!} + \frac{e^{\theta x} x^N}{N!},$$

where $0 < \theta < 1$. In particular, for $x = 1$, we have

$$e = 1 + 1 + \frac{1}{2!} + \cdots + \frac{1}{(N-1)!} + \frac{e^\theta}{N!},$$

for some θ where $0 < \theta < 1$. Suppose that e were rational, and $e = p/q$ where p, q are integers. If we take $N > \max\{q, 3\}$, then

$$(N-1)! \, p/q = (N-1)! \left(1 + 1 + \frac{1}{2!} + \cdots + \frac{1}{(N-1)!}\right) + e^\theta/N.$$

Now the left-hand side is an integer, and the first term on the right-hand side is an integer. This implies e^θ/N is an integer. Since $e^\theta < 3 < N$, this is impossible.

4.3   Power series expansion of a function

If the function f has derivatives of *all* orders in an interval containing a, then we can certainly write down the series

$$f(a) + \frac{f'(a)}{1!} h + \frac{f''(a)}{2!} h^2 + \cdots + \frac{f^{(n)}(a)}{n!} h^n + \cdots \tag{11}$$

(a power series in h) which we call the *Taylor series* of f at a. S_N as defined by (9) is therefore the Nth partial sum of (11). At this stage of course we make no claim concerning the convergence of (11), nor of any relation between (11) and $f(a + h)$. However, if $R_N \to 0$ as $N \to \infty$, then it follows from (10) that $S_N \to f(a + h)$ and so the series (11) is convergent and its sum is $f(a + h)$. We have therefore shown

PROPOSITION 4.6
Suppose that $f(x)$ has derivatives of all orders in an interval containing a, and that for certain values of h

$$R_N = \frac{h^N}{N!} f^{(N)}(a + \theta h) \to 0 \quad as \ N \to \infty.$$

Then for such h

$$f(a + h) = \sum_{n=0}^{\infty} \frac{f^{(n)}(a)}{n!} h^n,$$

i.e. the Taylor series of f at a converges with sum $f(a + h)$.

The special case when $a = 0$ is particularly useful. Explicitly we have

PROPOSITION 4.7
Suppose that $f(x)$ has derivatives of all orders in an interval containing the origin, and that for certain values of x

$$R_N = \frac{x^N}{N!}\, f^{(N)}(\theta x) \to 0 \quad as \; N \to \infty.$$

Then for such x the power series

$$f(0) + \frac{f'(0)}{1!}x + \frac{f''(0)}{2!}x^2 + \cdots + \frac{f^{(n)}(0)}{n!}x^n + \cdots \tag{12}$$

converges and $f(x)$ is its sum.

Example 6   Show that $\sin x = \mathrm{Sin}\, x$ for all values of x.

Solution   We recall that $\mathrm{Sin}\, x$ is defined by the convergent power series

$$\mathrm{Sin}\, x = x - \frac{x^3}{3!} + \frac{x^5}{5!} - \frac{x^7}{7!} + \cdots \quad \text{(Chapter 2, Ex. (xxii))}.$$

If $f(x) = \sin x$ then $f(x)$ has derivatives of all orders for all values of x, and $f^{(2n)}(x) = (-1)^n \sin x$, $f^{(2n+1)}(x) = (-1)^n \cos x$. It follows then that $f^{(2n)}(0) = 0$ and $f^{(2n+1)}(0) = (-1)^n$, and so the power series which defines $\mathrm{Sin}\, x$ is in fact the Taylor series for $\sin x$. To prove the required equality we need only show that $R_N \to 0$ as $N \to \infty$. But $|\sin(\theta x)| \leqslant 1$, $|\cos(\theta x)| \leqslant 1$, and so $R_N \leqslant |x|^N/N!$. From Chapter 2, Example 7 we see that for all x, $|x|^N/N! \to 0$ as $N \to \infty$, and so $R_N \to 0$ as $N \to \infty$.

EXERCISE (viii)
Show that $\cos x = \mathrm{Cos}\, x$ for all values of x.

Example 7   Find the Taylor series at the origin for $\tan^{-1} x$.

Solution   Put $f(x) = \tan^{-1} x$. Then $f'(x) = 1/(1 + x^2)$ and so

$$(1 + x^2)f'(x) = 1.$$

Now using Leibniz's rule (Proposition 3.13) for the nth derivative, we see that

$$(1 + x^2)f^{(n+1)}(x) + 2nxf^{(n)}(x) + n(n-1)f^{(n-1)}(x) = 0.$$

If we put $x = 0$ we obtain the recurrence relation

$$f^{(n+1)}(0) = -n(n-1)f^{(n-1)}(0).$$

Since $f(0) = 0$ and $f^{(1)}(0) = 1$, then $f^{(2n)}(0) = 0$ and $f^{(2n+1)}(0) = (-1)^n(2n)!$. Hence

$$\frac{f^{(2n+1)}(0)}{(2n+1)!}x^{2n+1} = \frac{(-1)^n x^{2n+1}}{2n+1}.$$

The Taylor series for $\tan^{-1} x$ is therefore

$$x - \frac{x^3}{3} + \frac{x^5}{5} - \frac{x^7}{7} + \cdots + \frac{(-1)^n x^{2n+1}}{2n+1} + \cdots .$$

Note that this series converges for $|x| \leqslant 1$ but not for $|x| > 1$.

EXERCISE (ix)
Obtain the Taylor series at the origin for

(a) $\sin^{-1} x$,   (b) $\cos^{-1} x$.

EXERCISE (x)
Obtain the Taylor series at the origin for $\log(1 + x)$, and show that this converges to $\log(1 + x)$ for $-1 < x \leqslant 1$. (It is fairly easy to prove convergence for $-\frac{1}{2} \leqslant x \leqslant 1$. To extend the range of convergence you may appeal to Proposition 2.23.)

4.4   Maxima and minima

The function $f(x)$ is said to have a *local maximum* at $x = \xi$ if ξ is the centre of some interval throughout which $f(x) \leqslant f(\xi)$. If the inequality is reversed then we say that $f(x)$ has a *local minimum* at $x = \xi$. For brevity we shall use the terms *maximum* and *minimum* instead. Since these terms refer to *local* properties of the function, it is quite possible for the value of $f(x)$ at a minimum to exceed the value of $f(x)$ at a maximum. Particular attention should also be paid to the values of $f(x)$ at the end points a, b of the range in question, as the values of $f(x)$ there may be greater or less than the values at a maximum or minimum respectively (see Fig. 9).

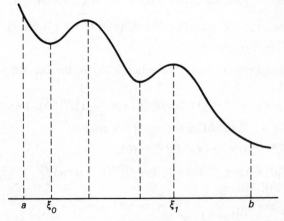

Figure 9

PROPOSITION 4.8

If $f(x)$ has a maximum or a minimum at $x = \xi$ and if $f(x)$ is differentiable at ξ then $f'(\xi) = 0$.

Proof

Suppose that $f'(\xi) \neq 0$. Then there exists $\delta > 0$ such that, for $0 < |h| < \delta$ the quotient $(f(\xi + h) - f(\xi))/h$ takes the same sign as $f'(\xi)$. This implies that the sign of $f(\xi + h) - f(\xi)$ changes with the sign of h. Therefore in every interval with ξ as centre $f(x) - f(\xi)$ takes values of both signs, and $f(x)$ cannot have a maximum or a minimum at ξ.

The vanishing of the derivative at ξ is therefore a necessary condition for a function to have a maximum or a minimum at ξ. This condition however is not sufficient. For example, with $f(x) = x^3$, we see that $f'(0) = 0$ but $x = 0$ is neither a maximum nor a minimum of this function. Points ξ for which $f'(\xi) = 0$ are called *stationary points*, and so points of maxima or minima are stationary points. It is particularly important for physical applications that we are able to distinguish between maxima, minima and other stationary points (these *other* points are known as *points of inflection*).

PROPOSITION 4.9

Suppose that $f'(\xi) = 0$, that $f^{(N)}(x)$ is the first derivative of f which does not vanish at ξ, and that $f^{(N)}(x)$ is continuous at ξ. Then if N is even, ξ is a maximum or a minimum according as $f^{(N)}(\xi)$ is negative or positive. If N is odd, then ξ is neither a maximum nor a minimum.

Proof

Taylor's expansion of $f(x)$ at $x = \xi$ tells us that

$$f(\xi + h) - f(\xi) = \frac{h^N}{N!} f^{(N)}(\xi + \theta h) \tag{13}$$

for all sufficiently small h. Since $f^{(N)}(\xi + \theta h) \to f^{(N)}(\xi)$ as $h \to 0$ then the right hand side of (13) takes the same sign as $h^N f^{(N)}(\xi)$ for sufficiently small $h, h \neq 0$. If N is even therefore the left-hand side of (13) takes the same sign as $f^{(N)}(\xi)$, while if N is odd then the left-hand side of (13) changes sign with h

EXERCISE (xi)

Find the stationary points of the following functions and classify them:

(a) $x^2 e^{-x}$,   (b) $x^2 + \cos x^2$,   (c) $x^7 + 14x^4$.

CHAPTER 5

Ordinary Differential Equations

5.1 Introduction

Suppose that y denotes a variable whose values are dependent on the independent variable x. A functional relation which involves the derivatives of y with respect to x is called an *ordinary differential equation*. The differential equation is said to be of *order n* if the highest derivative of y which appears is of order n, i.e. is $d^n y/dx^n$. For example,

$$\sin x \frac{dy}{dx} + y \cos x = e^x \sin x \tag{1}$$

and

$$\frac{dy}{dx} + 2xy = 4x \tag{2}$$

are differential equations of order 1, while

$$\frac{d^2 y}{dx^2} - 2 \frac{dy}{dx} + 10y = 0 \tag{3}$$

is of order 2. The aim in solving differential equations is to express y in terms of x by means of a relation of the form $y = \varphi(x)$ or $\psi(x, y) = 0$. In general more than one solution is possible, and our formula for y must therefore give all possible solutions.

The simplest form of differential equation is

$$\frac{dy}{dx} = f(x) \tag{4}$$

and the solution for this is

$$y = \int f(x)\, dx + C,$$

where C is an arbitrary constant. It is assumed that the reader is familiar with the integrals of elementary functions and the techniques of integration involving changes of variable, partial fractions and integration by parts. Our aim is to solve equations of a rather more complicated nature.

EXERCISE (i)
Solve the differential equation (4) when $f(x)$ is

(a) $\sin x \cos x$, (b) $(4 - x^2)^{-1/2}$, (c) $x^{-1} \log x$,
(d) $\sin 5x \sin 7x$, (e) $(x^2 + 2x + 5)^{-1}$, (f) 3^x, (g) xe^x,
(h) $x^3 \log x$, (i) $\tan^{-1} x$, (j) $(3x + 4)/((2x + 1)(x + 3)^2)$.

5.2 First order equations

These equations are of the form

$$\frac{dy}{dx} = f(x, y),\tag{5}$$

and we solve this when the function $f(x, y)$ is of certain special forms.
First we consider the case when $f(x, y) = g(x)/h(y)$. In this case we say that the *variables separate*. The equation (5) then takes the form

$$h(y)\frac{dy}{dx} = g(x),\tag{6}$$

and the complete solution is given by

$$\int h(y)\, dy = \int g(x)\, dx + C.$$

Example 1 Solve $y^2 (x - 1)\, dy/dx + x^2(y + 1) = 0$.

Solution This may be rewritten in the form

$$\frac{y^2}{y + 1}\frac{dy}{dx} = \frac{-x^2}{x - 1}$$

and the variables separate. Thus the solution is

$$\int \frac{y^2}{y + 1}\, dy = - \int \frac{x^2}{x - 1}\, dx + C,$$

which on integration gives

$$\tfrac{1}{2}(y - 1)^2 + \log(y + 1) = -\tfrac{1}{2}(x + 1)^2 - \log(x - 1) + C.$$

EXERCISE (ii)
Solve

(a) $\dfrac{dy}{dx} = 2e^y$, (b) $x^2 \dfrac{dy}{dx} = \cos^2 y$, (c) $e^x \dfrac{dy}{dx} = xe^y$,

(d) $\dfrac{dy}{dx} = x^2 + e^y + x^2 e^y + 1$, (e) $x(3 - y)\dfrac{dy}{dx} = 4y$.

If the function $f(x, y)$ of (5) takes the form $g(y/x)$, then the equation takes the form

$$\frac{dy}{dx} = g\left(\frac{y}{x}\right) \tag{7}$$

and we say that it is *homogeneous*. In this case we make the substitution $t = y/x$. Since $dy/dx = t + x\, dt/dx$, then (7) becomes

$$t + x\frac{dt}{dx} = g(t)$$

and we see that the variables separate.

Example 2 Solve $(x^2 + y^2)\, dy/dx - 2xy = 0$.

Solution This can be written in the form $dy/dx = 2(y/x)(1 + (y/x)^2)^{-1}$, and so is homogeneous. The substitution $t = y/x$ yields

$$t + x\frac{dt}{dx} = \frac{2t}{1 + t^2},$$

i.e.

$$x\frac{dt}{dx} = \frac{t - t^3}{1 + t^2}.$$

Hence

$$\int \frac{1 + t^2}{t - t^3}\, dt = \int \frac{dx}{x} + c,$$

and by using partial fractions we obtain

$$\log \frac{t}{1 - t^2} = \log x + c$$

i.e.

$$Ct = x(1 - t^2) \qquad \text{or} \qquad Cy = x^2 - y^2.$$

Example 3 Solve $dy/dx = (2x + y + 5)/(x + 8y - 5)$.

Solution This equation is not homogeneous as it stands. However, if we make the substitution $x = X + a, y = Y + b$ and choose a and b so that $2a + b + 5 = 0, a + 8b - 5 = 0$ (i.e. $a = -3, b = 1$). Then the equation becomes

$$\frac{dY}{dX} = \frac{2X + Y}{X + 8Y}$$

which is homogeneous. Now put $T = Y/X$. Then

$$T + X \frac{dT}{dX} = \frac{2 + T}{1 + 8T},$$

i.e.

$$X \frac{dT}{dX} = \frac{2(1 - 4T^2)}{1 + 8T}.$$

Therefore

$$\log X = \frac{1}{2} \int \frac{1 + 8T}{1 - 4T^2} \, dT$$

$$= -\tfrac{5}{8} \log (1 - 2T) - \tfrac{3}{8} \log (1 + 2T) + c,$$

or

$$\tfrac{5}{8} \log (X - 2Y) + \tfrac{3}{8} \log (X + 2Y) = c.$$

In terms of x and y, this can be written

$$(x - 2y + 5)^5 (x + 2y + 1)^3 = C,$$

where C is an arbitrary constant.

EXERCISE (iii)
Solve

(a) $\dfrac{dy}{dx} = \dfrac{y - x}{x + y}$,   (b) $xy \dfrac{dy}{dx} = x^2 + 4xy + 3y^2$,

(c) $x^2 \dfrac{dy}{dx} = y^2 + 2xy + x^2$,   (d) $\dfrac{dy}{dx} = \dfrac{x + y - 3}{5x - 3y + 1}$,

(e) $\dfrac{dy}{dx} = \dfrac{12x + y + 8}{x + 3y - 11}$.

If the equation (5) takes the form

$$\frac{dy}{dx} + yg(x) = h(x) \tag{8}$$

then we say that is a (first order) *linear equation*. We treat such equations by multiplying by a term

$$e^{\int g(x)\,dx}$$

which is called an *integrating factor*. The equation (8) then becomes

$$e^{\int g(x)\,dx}\,\frac{dy}{dx} + y\,g(x)\,e^{\int g(x)\,dx} = h(x)\,e^{\int g(x)\,dx}$$

or

$$\frac{d}{dx}\left(y\,e^{\int g(x)\,dx}\right) = h(x)\,e^{\int g(x)\,dx}.$$

This can now be integrated to give

$$y\,e^{\int g(x)\,dx} = \int h(x)\,e^{\int g(x)\,dx}\,dx + c,$$

which yields a formula for y.

Example 4   Solve the equation (2).

Solution   Equation (2) is in the form (8) with $g(x) = 2x$. Since $\int 2x\,dx = x^2$ the integrating factor is e^{x^2} and multiplying through we have

$$e^{x^2}\,\frac{dy}{dx} + 2x\,e^{x^2}\,y = 4x\,e^{x^2}$$

or

$$\frac{d}{dx}\left(e^{x^2}y\right) = 4x\,e^{x^2}.$$

Hence

$$e^{x^2}y = \int 4x\,e^{x^2}\,dx + c = 2\,e^{x^2} + c$$

and so

$$y = 2 + c\,e^{-x^2}.$$

Example 5   Solve $x\,dy/dx - 2y = (x - 2)e^x$.

Solution   We recognize this equation as linear if we rewrite it in the form

$$\frac{dy}{dx} - \frac{2y}{x} = \frac{(x - 2)}{x}\,e^x.$$

Now

$$\int \frac{-2\,dx}{x} = -2\log x = \log x^{-2},$$

and so the integrating factor is $e^{\log x^{-2}}$, i.e. x^{-2}. Hence

$$x^{-2}\,\frac{dy}{dx} - 2x^{-3}y = (x - 2)x^{-3}e^x$$

and so

$$\frac{d}{dx}(x^{-2}y) = (x-2)x^{-3}e^x.$$

Therefore

$$x^{-2}y = \int (x-2)x^{-3}e^x\,dx + c$$

$$= x^{-2}e^x + c,$$

which yields

$$y = e^x + cx^2.$$

EXERCISE (iv)
Solve

(a) $\dfrac{dy}{dx} - y\tan x = e^x \sec x,$ (b) $(x+1)\dfrac{dy}{dx} + y = x - 1,$

(c) $(1-x^2)\dfrac{dy}{dx} + 2xy = x^2,$ (d) $(x-1)\dfrac{dy}{dx} + xy = e^{-x}\cos x,$

(e) $(1+x^3)\dfrac{dy}{dx} = x^2y.$

Finally, we consider the case when the first order differential equation may be expressed in the form

$$g(x,y)\frac{dy}{dx} + h(x,y) = 0, \tag{9}$$

where $\partial g/\partial x = \partial h/\partial y$ (we refer the reader to §6.2 for explanation of this notation). When this occurs we say that (9) is *exact*, and in this case there is a function $F(x,y)$ such that the left hand side of (9) is equal to $(d/dx)F(x,y)$. The solution of (9) therefore is given by

$$F(x,y) = c,$$

where c is an arbitrary constant. (The student who has not previously encountered the symbols for partial differentiation: $\partial g/\partial x$, $\partial h/\partial y$, will find an explanation in Chapter 6.)

Example 6   Solve $(3x^4y^2 - x^2)\,dy/dx + (4x^3y^3 - 2xy) = 0.$

Solution   Note that

$$\frac{\partial}{\partial x}(3x^4y^2 - x^2) = 12x^3y^2 - 2x = \frac{\partial}{\partial y}(4x^3y^3 - 2xy)$$

and so our equation is exact. The problem of finding $F(x, y)$ is fairly simple since the left hand side of the equation may be rewritten in the form

$$\left(3x^4 y^2 \frac{dy}{dx} + 4x^3 y^3\right) - \left(x^2 \frac{dy}{dx} + 2xy\right)$$

or

$$\frac{d}{dx}(x^4 y^3) - \frac{d}{dx}(x^2 y)$$

and so

$$F(x, y) = x^4 y^3 - x^2 y.$$

The solution therefore is

$$x^4 y^3 - x^2 y = c.$$

The equation

$$xy \frac{dy}{dx} + x^2 + y^2 + x = 0$$

is not exact. However, multiplication by x gives

$$x^2 y \frac{dy}{dx} + x^3 + xy^2 + x^2 = 0$$

which is exact and easily solved (the solution being $\frac{1}{2}x^2 y^2 + \frac{1}{4}x^4 + \frac{1}{3}x^3 = c$). Thus it is convenient to introduce an integrating factor, in this case x. In our previous treatment of first order linear equations the effect of multiplying by an integrating factor was in fact to produce an exact equation. In the present case equally, a non-exact equation (9) may be rendered exact by multiplication by a suitable integrating factor. General guidelines are the following:

I if $(\partial/\partial y)h(x, y) - (\partial/\partial x)g(x, y))/g(x, y)$ is a function of x alone, say $\varphi(x)$, then $e^{\int \varphi(x)\,dx}$ is a suitable integrating factor;

II if $((\partial/\partial x)g(x, y) - (\partial/\partial y)h(x, y))/h(x, y)$ is a function of y alone, say $\psi(y)$, then $e^{\int \psi(y)\,dy}$ is a suitable integrating factor.

Example 7   Solve $\cos x \, dy/dx + y \sin x = (y \sec x)^{1/2}$.

Solution   This equation is not exact, and regrettably, the straightforward approach will not reveal the required integrating factor. Some slight subtlety is called for. Multiplying through by $y^{-1/2}$ we obtain.

$$y^{-1/2} \cos x \frac{dy}{dx} + y^{1/2} \sin x - (\sec x)^{1/2} = 0$$

and considering this instead we see that

$$\frac{\frac{\partial}{\partial y}(y^{1/2}\sin x - (\sec x)^{1/2}) - \frac{\partial}{\partial x}(y^{-1/2}\cos x)}{y^{-1/2}\cos x} = \frac{3}{2}\tan x.$$

An integrating factor is $e^{\int 3/2\,\tan x\,dx}$, i.e. $(\cos x)^{-3/2}$. Multiplying through by this gives

$$y^{-1/2}(\cos x)^{-1/2}\frac{dy}{dx} + y^{1/2}\sin x\,(\cos x)^{-3/2} - \sec^2 x = 0$$

or

$$\frac{d}{dx}(2y^{1/2}(\cos x)^{-1/2} - \tan x) = 0.$$

Therefore

$$2y^{1/2}(\cos x)^{-1/2} - \tan x = c$$

and so

$$y = \tfrac{1}{4}\cos x\,(c + \tan x)^2.$$

EXERCISE (v)
Solve

(a) $(2y^3 - x)\dfrac{dy}{dx} + y = 0$,   (b) $y^2\sin x\dfrac{dy}{dx} + y^3\cos x = 1$,

(c) $(3y + x^2)\dfrac{dy}{dx} + yx = 0$,   (d) $(3xy^2 - 1)\dfrac{dy}{dx} + 5y^2 + 3y^3 = xy^2$.

5.3   Linear equations with constant coefficients

So far we have dealt only with first order equations. Among equations of higher order there is one class whose solutions are fairly easily obtained. An equation of the form

$$a_n(x)\frac{d^n y}{dx^n} + a_{n-1}(x)\frac{d^{n-1}y}{dx^{n-1}} + \cdots + a_1(x)\frac{dy}{dx} + a_0(x)y = f(x) \quad (10)$$

where the $a_0(x), a_1(x), \ldots, a_n(x), f(x)$ are functions of x is called a *linear differential equation*. Equation (8) is such an equation of the first order. The class of equations we shall study here consists of those linear equations where the coefficients $a_0(x), \ldots, a_n(x)$ are *constants*, i.e. equations of the form

$$a_n\frac{d^n y}{dx^n} + a_{n-1}\frac{d^{n-1}y}{dx^{n-1}} + \cdots + a_1\frac{dy}{dx} + a_0 y = f(x), \quad (11)$$

where $a_0, a_1, \ldots, a_n$ are constants (possibly complex numbers), $a_n \neq 0$.

We shall use the symbol D^n to denote the operation of differentiating n times, and D^n is called an *operator*. Thus for example

$$Dg(x) = \frac{d}{dx} g(x), \qquad D^2 g(x) = \frac{d^2}{dx^2} g(x), \ldots,$$

$$D^n g(x) = \frac{d^n}{dx^n} g(x)$$

if $g(x)$ is a function with nth derivative. In this notation the equation (11) is more conveniently written

$$(a_n D^n + a_{n-1} D^{n-1} + \cdots + a_1 D + a_0) y = f(x). \tag{12}$$

If we denote by $F(D)$ the polynomial in D of degree n which occurs in (12) then the equation becomes

$$F(D) y = f(x). \tag{13}$$

EXERCISE (vi)
Suppose that $F(D)$ and $G(D)$ are polynomials in the operator D, that $g(x)$ and $h(x)$ are functions which are differentiable a suitable number of times and that c is a constant. Show that

(a) $F(D)(g(x) + h(x)) = F(D) g(x) + F(D) h(x)$;

(b) $F(D)(cg(x)) = cF(D) g(x)$;

(c) $(F(D) + G(D)) g(x) = F(D) g(x) + G(D) g(x)$;

(d) $(F(D) G(D)) g(x) = F(D)(G(D) g(x)) = G(D)(F(D) g(x))$.

(Note that by "polynomial" we mean "polynomial with real or complex coefficients". This is important.)

The polynomial $F(D)$ of degree n has n (not necessarily distinct) roots. If the distinct roots are $m_1, m_2, \ldots, m_s$ say, of multiplicities $r_1, r_2, \ldots, r_s$ respectively then

$$F(D) = a_n (D - m_1)^{r_1} (D - m_2)^{r_2} \ldots (D - m_s)^{r_s}$$

and, in view of Exercise (vi) (d), (13) may be expressed in the form

$$a_n (D - m_1)^{r_1} (D - m_2)^{r_2} \ldots (D - m_s)^{r_s} y = f(x) \tag{14}$$

We remind ourselves that our aim is to find *all* solutions of (13). Suppose then that somehow we have managed to obtain *one* such solution, say $h(x)$. Such a solution is called a *particular integral* of (13). If $g(x)$ is any other solution then by Exercise (vi) (a)

$$F(D)(g(x) - h(x)) = F(D) g(x) - F(D) h(x) = f(x) - f(x) = 0.$$

In other words, $g(x) - h(x)$ is a solution of the *reduced equation*

$$F(D) y = 0$$

or

$$a_n(D - m_1)^{r_1}(D - m_2)^{r_2} \dots (D - m_s)^{r_s}y = 0 \qquad (15)$$

On the other hand, if $j(x)$ is any solution of (15), then $h(x) + j(x)$ is a solution of (13). Our problem therefore is split into two parts: firstly we must find a particular integral of (13), and secondly we must find the general solution of the reduced equation (15) (which is called the *complementary function* of (13)). The general solution of (13) is then given by the formula

  particular integral + complementary function.

To obtain the complementary function we shall require the following

PROPOSITION 5.1

$$(D - m)^r y = e^{mx}D^r(y \, e^{-mx}).$$

Proof
The proof is by induction on r. For $r = 1$,

$$e^{mx}D(y \, e^{-mx}) = e^{mx}(e^{-mx}Dy - ym \, e^{-mx}) = Dy - my.$$

Assuming the proposition is valid for $r = k$,

$$\begin{aligned}
(D - m)^{k+1}y &= (D - m)(e^{mx}D^k(y \, e^{-mx})) \\
&= D(e^{mx}D^k(y \, e^{-mx})) - m \, e^{mx}D^k(y \, e^{-mx}) \\
&= e^{mx}D^{k+1}(y \, e^{-mx}) + m \, e^{mx}D^k(y \, e^{-mx}) \\
&\quad - m \, e^{mx}D^k(y \, e^{-mx}) \\
&= e^{mx}D^{k+1}(y \, e^{-mx}).
\end{aligned}$$

We have shown that the formula is valid for $r = k + 1$, and it follows therefore that it is valid for all positive integers r.

We deduce from this proposition that if y is any solution of

$$(D - m)^r y = 0 \qquad (16)$$

then y satisfies

$$D^r(y \, e^{-mx}) = 0$$

and so a solution of (16) is given by

$$y \, e^{-mx} = C_1 + C_2 x + \dots + C_r x^{r-1},$$

or

$$y = e^{mx}P(x)$$

where $P(x)$ is an arbitrary polynomial of degree $r - 1$.
  Since the factors of the operators of (15) commute, then we see that

$$e^{m_1 x}P_1(x), \qquad e^{m_2 x}P_2(x), \qquad \dots, \qquad e^{m_s x}P_s(x),$$

where $P_i(x)$ are arbitrary polynomials of degrees $r_i - 1$ $(i = 1, \ldots, s)$, are also solutions of (15). By linearity therefore (Ex. (vi)(a))

$$e^{m_1 x}P_1(x) + e^{m_2 x}P_2(x) + \cdots + e^{m_s x}P_s(x) \tag{17}$$

satisfies (15). *The expression* (17) *is the complementary function of the equation* (13). To see this we must prove

PROPOSITION 5.2
Every solution of the equation (15) *is of the form* (17).

Proof
The proof is by induction on the degree n of $F(D)$. The proposition is true for $n = 1$ since any solution of $(D - m)y = 0$ also satisfies $D(y\,e^{-mx}) = 0$, and so $y = C\,e^{mx}$ (Chapter 4, Example 1). Now assume that the proposition holds for polynomials of degree n, and suppose that $G(D) = (D - m)F(D) = (D - m)(D - m_1)^{r_1} \ldots (D - m_s)^{r_s}$. Then if y satisfies

$$G(D)y = 0$$

then

$$F(D)\{(D - m)y\} = 0,$$

and by the induction hypothesis

$$(D - m)y = e^{m_1 x}P_1(x) + \cdots + e^{m_s x}P_s(x).$$

From Proposition 5.1 therefore

$$D(y\,e^{-mx}) = e^{(m_1 - m)x}P_1(x) + \cdots + e^{(m_s - m)x}P_s(x),$$

and so

$$y = e^{mx}\left\{ \int e^{(m_1 - m)}P_1(x)\,dx + \cdots + \int e^{(m_s - m)x}P_s(x)\,dx \right\}.$$

We note that this is of the required form when we observe that

$$\int e^{(m_i - m)x}P_i(x)\,dx = e^{(m_i - m)x}Q_i(x),$$

where $Q_i(x)$ is a polynomial of the same degree as $P_i(x)$ when $m_i \neq m$, and is of degree one greater when $m_i = m$.

Example 8   Verify that $y = x^3 - 2$ is a solution of $(D^3 - 3D + 2)y = 2x^3 - 9x^2 + 2$, and find the general solution.

Solution $(D^3 - 3D + 2)(x^3 - 2) = 6 - 9x^2 + 2x^3 - 4 = 2x^3 - 9x^2 + 2$. Hence $y = x^3 - 2$ is a particular integral of the differential equation. To find the complementary function we solve

$$m^3 - 3m + 2 = 0$$

(this is called the *auxiliary equation*), and find that the roots are 1, 1 and −2. Therefore, using the formula (17), the complementary function is

$$e^x(Ax + B) + C\,e^{-2x},$$

and hence the required general solution is

$$y = x^3 - 2 + e^x(Ax + B) + C\,e^{-2x}.$$

where A, B and C are arbitrary constants.

It is quite possible that the auxiliary equation may turn out to have complex roots. In this case the solution is more appropriately expressed in terms of circular functions.

Example 9   Solve $(D^3 - 2D^2 + 10D)y = 0$.

Solution   The equation is already in reduced form. Its auxiliary equation $m^3 - 2m^2 + 10m = 0$ has roots 0, $1 + 3i$ and $1 - 3i$, and so the complementary function (which in this case happens to be the general solution) is

$$A + B\,e^{(1+3i)x} + C\,e^{(1-3i)x}.$$

Using the formula (29) of Chapter 2, this may be expressed in the form

$$A + e^x(B(\cos 3x + i \sin 3x) + C(\cos 3x - i \sin 3x)),$$

or

$$A + e^x((B + C)\cos 3x + (iB - iC)\sin 3x).$$

Since A, B and C are arbitrary constants (which could be complex) we may write the solution in the form

$$A + e^x(D \cos 3x + E \sin 3x)$$

where A, D and E are arbitrary constants.

EXERCISE (vii)
Find particular integrals for the following equations and solve them completely:

(a) $(D^2 - 2D + 1)y = 1$,   (b) $(D^3 - 2D^2 - 4D + 8)y = 0$,

(c) $(D^3 + 4D)y = 4$,   (d) $(D^3 - 3D^2 + 3D - 1)y = 0$,

(e) $(D^2 - 2)y = 4x$.

We now know how to find the complementary function and so the next problem is to find a particular integral. The particular integrals required in Ex. (vii) are not difficult to find, and can be found by observation. Techniques for finding a suitable particular integral in certain cases are given below. We shall use the notation

$$\frac{1}{F(D)} f(x) \tag{18}$$

to denote a particular integral of $F(D)y = f(x)$. Note that the symbol (18) does not denote a unique function, but in fact may be used to denote *any* function which satisfies our differential equation. For example we may write (cf. Example 8)

$$\frac{1}{D^3 - 3D + 2}(2x^3 - 9x^2 + 2) = x^3 - 2,$$

but equally true is

$$\frac{1}{D^3 - 3D + 2}(2x^3 - 9x^2 + 2) = x^3 - 2 + e^x + 2e^{-2x}$$

It should be noted that the symbol $(1/D)f(x)$ represents "the integral of $f(x)$", which therefore agrees with the notion of integration being the inverse operation to differentiation. The student should guard against misuse of notation: the symbols

$$\frac{f(x)}{F(D)}, \qquad f(x)\frac{1}{F(D)}$$

are not defined at all, and should not be used instead of (18).

PROPOSITION 5.3

$$G(D)\left(\frac{1}{F(D)}f(x)\right) = \frac{1}{F(D)}(G(D)f(x)).$$

Proof

Since

$$F(D)\left(G(D)\left(\frac{1}{F(D)}f(x)\right)\right) = G(D)\left(F(D)\left(\frac{1}{F(D)}f(x)\right)\right) = G(D)f(x),$$

then the result follows.

LEMMA 5.4

(i) $F(D)\, e^{\alpha x} = e^{\alpha x}F(\alpha)$;

(ii) $F(D)(e^{\alpha x}V(x)) = e^{\alpha x}(F(D + \alpha)\, V(x))$;

(iii) $F(D)(xV(x)) = x(F(D)\, V(x)) + F'(D)\, V(x)$;

where α is any constant, $V(x)$ is any function of x with enough derivatives, and $F'(D)$ is the derivative with respect to D of $F(D)$.

Proof

(i) Since $D\, e^{\alpha x} = \alpha\, e^{\alpha x},\ D^2\, e^{\alpha x} = \alpha^2\, e^{\alpha x}, \ldots, D^n\, e^{\alpha n} = \alpha^n\, e^{\alpha x}$, then

$$F(D) e^{\alpha x} = (a_n D^n + a_{n-1} D^{n-1} + \cdots + a_1 D + a_0) e^{\alpha x}$$
$$= (a_n \alpha^n + a_{n-1} \alpha^{n-1} + \cdots + a_1 \alpha + a_0) e^{\alpha x}$$
$$= F(\alpha) e^{\alpha x}.$$

(ii) By Leibniz's rule for the higher derivatives of a product (Proposition 3.13) we have

$$D^m(e^{\alpha x} V(x)) = \sum_{r=0}^{m} \binom{m}{r} (D^r e^{\alpha x})(D^{m-r} V(x))$$

$$= e^{\alpha x} \sum_{r=0}^{m} \binom{m}{r} \alpha^r D^{m-r} V(x)$$

$$= e^{\alpha x}((D + \alpha)^m V(x)), \quad \text{for } m = 1, 2, 3, \ldots.$$

Therefore

$$F(D)(e^{\alpha x} V(x)) = (a_n D^n + a_{n-1} D^{n-1} + \cdots + a_1 D + a_0)(e^{\alpha x} V(x))$$
$$= e^{\alpha x}(a_n(D + \alpha)^n + a_{n-1}(D + \alpha)^{n-1} + \cdots$$
$$+ a_1(D + \alpha) + a_0) V(x)$$
$$= e^{\alpha x}(F(D + \alpha) V(x)).$$

(iii) Leibniz's rule applied to the function $xV(x)$ gives us

$$D^m(xV(x)) = xD^m V(x) + mD^{m-1} V(x), \quad \text{for } m = 1, 2, 3, \ldots.$$

Hence

$$F(D)(xV(x)) = (a_n D^n + a_{n-1} D^{n-1} + \cdots + a_1 D + a_0)(xV(x))$$
$$= a_n(xD^n V(x) + nD^{n-1} V(x)) + a_{n-1}(xD^{n-1} V(x)$$
$$+ (n-1)D^{n-2} V(x)) + \cdots + a_1(xDV(x) + V(x))$$
$$+ a_0 xV(x)$$
$$= x(a_n D^n + a_{n-1} D^{n-1} + \cdots + a_1 D + a_0) V(x)$$
$$+ (a_n n D^{n-1} + a_{n-1}(n-1)D^{n-2} + \cdots + a_1) V(x)$$
$$= x(F(D) V(x)) + F'(D) V(x).$$

LEMMA 5.5
Suppose that in the polynomial $F(D)$ each term of the form $a_{2m} D^{2m}$ is replaced by $a_{2m}(-\alpha^2)^m$, and each term of the form $a_{2m+1} D^{2m+1}$ is replaced by $a_{2m+1}(-\alpha^2)^m D$. If $aD + b$ denotes the resulting polynomial of degree 1, then

$$F(D) \cos \alpha x = (aD + b) \cos \alpha x,$$
$$F(D) \sin \alpha x = (aD + b) \sin \alpha x.$$

Proof
The result follows from linearity and the fact that

$$a_{2m}D^{2m}\cos\alpha x = a_{2m}(-\alpha^2)^m \cos\alpha x,$$
$$a_{2m+1}D^{2m+1}\cos\alpha x = a_{2m+1}(-\alpha^2)^m(D\cos\alpha x),$$

and similarly for $\sin\alpha x$.

The following propositions give formulae for the particular integral (18) when $f(x)$ is of certain standard forms.

PROPOSITION 5.6
If $F(\alpha) \neq 0$ then $(1/F(D)) e^{\alpha x} = (1/F(\alpha)) e^{\alpha x}$.

Proof
All we have to prove is that $F(D)(1/F(\alpha)) e^{\alpha x} = e^{\alpha x}$. This follows from Lemma 5.4 (i) since the left-hand side is equal to $(1/F(\alpha))(F(D) e^{\alpha x})$.

PROPOSITION 5.7
If $F(\alpha) = 0$ and we factorize $F(D) = (D - \alpha)^r G(D)$ where $G(\alpha) \neq 0$, then

$$\frac{1}{F(D)} e^{\alpha x} = \frac{1}{G(\alpha)} e^{\alpha x} \frac{x^r}{r!} \,.$$

Proof

Since

$$F(D)\left\{ \frac{1}{G(\alpha)} e^{\alpha x} \frac{x^r}{r!} \right\} = \frac{1}{G(\alpha)} G(D)\left\{ (D-\alpha)^r\left(e^{\alpha x} \frac{x^r}{r!} \right) \right\}$$

$$= \frac{1}{G(\alpha)} G(D)\left\{ e^{\alpha x}\left(D^r \frac{x^r}{r!} \right) \right\} \quad \text{(Lemma 5.4(ii))}$$

$$= \frac{1}{G(\alpha)} G(D)(e^{\alpha x}) = e^{\alpha x}, \quad \text{(Lemma 5.4 (i))}$$

the result follows.

Example 10   Solve $(D^2 + 4D + 3)y = 48\cosh 3x$.

Solution   The auxiliary equation has roots -3 and -1, and so the complementary function is $A e^{-3x} + B e^{-x}$ where A and B are arbitrary constants. To find the particular integral, we note that

$$\frac{1}{F(D)} \{f(x) + g(x)\} = \frac{1}{F(D)} f(x) + \frac{1}{F(D)} g(x) \tag{19}$$

and so

$$\frac{1}{D^2 + 4D + 3} 48 \cosh 3x = \frac{1}{D^2 + 4D + 3} (24\ e^{3x})$$

$$+ \frac{1}{D^2 + 4D + 3} (24\ e^{-3x}).$$

Now

$$\frac{1}{D^2 + 4D + 3} (24\ e^{3x}) = 24 \frac{1}{3^2 + 4.3 + 3} e^{3x} = e^{3x},$$

and

$$\frac{1}{D^2 + 4D + 3} (24\ e^{-3x}) = 24 \frac{1}{(D+3)(D+1)} e^{-3x}$$

(then use Proposition 5.7)

$$= 24 \frac{1}{(-3+1)} e^{-3x} \frac{x}{1!} = -12x\ e^{-3x}.$$

Therefore the required particular integral is $e^{3x} - 12x\ e^{-3x}$, and the complete solution is $y = e^{3x} - 12x\ e^{-3x} + A\ e^{-3x} + B\ e^{-x}$.

EXERCISE (viii)
Solve

(a) $(D^2 + 2)y = 3\ e^x + 4,$ (b) $(D^2 - 4D - 4)y = e^{5x},$

(c) $(D^2 - 6D + 9)y = 2\ e^{3x},$ (d) $(D^3 - 3D + 2)y = 6e^x + 12\ e^{-x}.$

PROPOSITION 5.8
If $aD + b$ is the polynomial of Lemma 5.5, *and a, b are not both zero, then*

$$\frac{1}{F(D)} \sin \alpha x = \frac{1}{aD + b} \sin \alpha x,$$

$$\frac{1}{F(D)} \cos \alpha x = \frac{1}{aD + b} \cos \alpha x.$$

Proof

$$F(D) \left(\frac{1}{aD + b} \sin \alpha x \right) = \frac{1}{aD + b} (F(D) \sin \alpha x) \quad \text{(Proposition 5.3)}$$

$$= \frac{1}{aD + b} ((aD + b) \sin \alpha x) = \sin \alpha x.$$

Similarly for $\cos \alpha x$.

Example 11   Find a particular integral of $(D^3 - 3D^2 + 5D - 11)y = \cos 2x$.

Solution

$$\frac{1}{D^3 - 3D^2 + 5D - 11} \cos 2x$$

$$= \frac{1}{(-2^2)D - 3(-2^2) + 5D - 11} \cos 2x$$

$$= \frac{1}{D + 1} \cos 2x = (D - 1)\left(\frac{1}{(D - 1)(D + 1)} \cos 2x\right)$$

$$= (D - 1)\left(\frac{1}{(-2^2) - 1} \cos 2x\right) = -\tfrac{1}{5}(D - 1) \cos 2x$$

$$= \tfrac{1}{5}(\cos 2x + 2 \sin 2x).$$

When $aD + b = 0$ then the technique of Example 11 is no longer applicable. The method then is to treat $\cos \alpha x$ and $\sin \alpha x$ as the real part and imaginary part respectively of $e^{i\alpha x}$, and proceed as in Example 10.

Example 12   Find a particular integral of $(D^3 + D^2 + 4D + 4)y = \sin 2x$.

Solution   Since $(-2^2)D + (-2^2) + 4D + 4 = 0$ then the method of Example 11 will not work. Instead we find a particular integral of $(D^3 + D^2 + 4D + 4)y = e^{2ix}$ and take its imaginary part. Thus

$$\frac{1}{D^3 + D^2 + 4D + 4} e^{2ix} = \frac{1}{(D - 2i)(D + 2i)(D + 1)} e^{2ix}$$

$$= \frac{1}{(2i + 2i)(2i + 1)} e^{2ix} x \quad \text{(Proposition 5.7)}$$

$$= \frac{-1}{20}(2 + i)(\cos 2x + i \sin 2x) x$$

$$= \frac{-x}{20}\{2 \cos 2x - \sin 2x + i(\cos 2x$$

$$+ 2 \sin 2x)\}.$$

The required particular integral is therefore

$$\frac{-x}{20}(\cos 2x + 2 \sin 2x).$$

EXERCISE (ix)
Solve

(a) $(D^2 - 2)y = e^{2x} - \sin 2x$,   (b) $(D^2 + 1)y = \cos x$,

(c) $(D^3 + 6D^2 + 11D + 6)y = 390 \sin 3x$,

(d) $(D^4 + 8D^2 + 16)y = \sin 2x$.

PROPOSITION 5.9
Suppose that $F(D) = D^r G(D)$ where $G(0) \neq 0$ (i.e. $G(D)$ has a non-zero constant term) and that $f(x)$ is a polynomial of degree m. Then $(1/F(D))f(x)$ is a polynomial of degree $m + r$.

Proof
Let $F(D) = a_n D^n + a_{n-1} D^{n-1} + \cdots + a_r D^r$ and $f(x) = b_m x^m + b_{m-1} x^{m-1} + \cdots + b_1 x + b_0$ where $a_r \neq 0$ and $b_m \neq 0$. The problem is equivalent to showing that there exist $c_0, c_1, \ldots, c_{m+r}$ such that

$$(a_n D^n + a_{n-1} D^{n-1} + \cdots + a_r D^r)(c_0 + c_1 x + \cdots + c_{m+r} x^{m+r})$$
$$= f(x).$$

Differentiating out and equating the coefficients of powers of x, we get

$$b_m = a_r \frac{(m+r)!}{m!} c_{m+r},$$

$$b_{m-1} = a_r \frac{(m+r-1)!}{(m-1)!} c_{m+r-1} + a_{r+1} \frac{(m+r)!}{(m-1)!} c_{m+r},$$

$$b_{m-2} = a_r \frac{(m+r-2)!}{(m-2)!} c_{m+r-2} + a_{r+1} \frac{(m+r-1)!}{(m-2)!} c_{m+r-1}$$

$$+ a_{r+2} \frac{(m+r)!}{(m-2)!} c_{m+r}$$

and so on. This process yields $c_{m+r}, c_{m+r-1}, \ldots, c_1, c_0$.

Example 13   Find a particular integral of $(D^4 + 4D^2)y = 48x^2 + 8$.

Solution   Here $r = 2$ and $m = 2$, so it is appropriate to consider as particular integral a polynomial $ax^4 + bx^3 + cx^2 + dx + e$ of degree 4. Since $(D^4 + 4D^2)(ax^4 + bx^3 + cx^2 + dx + e) = 48ax^2 + 24bx + 8c + 24a$, by equating coefficients we see that

$48a = 48$,

$24b = 0$,

$8c + 24a = 8$.

Therefore $a = 1, b = 0, c = -2$ and no restriction is placed on d and e. In particular we can take $d = e = 0$ and obtain the particular integral $x^4 - 2x^2$.

EXERCISE (x)
Solve

(a) $(D^2 + 4D - 5)y = 5x^2 + 2x + 5$,   (b) $(D^2 + 4)y = \cos 2x + 8x^3$,

(c) $(D^3 + 2D^2 + D)y = 3x^2 + 10x + 2 - 4e^{-x}$,

(d) $(D^3 - D^2 - D + 1)y = e^{2x} + 2x^3 - 6x^2 - 9x + 3$.

PROPOSITION 5.10

(i) $\dfrac{1}{F(D)}(e^{\alpha x} V(x)) = e^{\alpha x}\left(\dfrac{1}{F(D + \alpha)} V(x)\right)$,

(ii) $\dfrac{1}{F(D)}(x V(x)) = x\left(\dfrac{1}{F(D)} V(x)\right) - F'(D)\left(\dfrac{1}{(F(D))^2} V(x)\right)$.

Proof
The proof consists of straightforward applications of Lemma 5.4 (ii) and (iii) respectively to the right-hand sides of these equations.

EXERCISE (xi)
Solve

(a) $(D^3 - 6D^2 + 12D - 8)y = x^2 e^{2x}$,

(b) $(D^2 + 4D + 5)y = 10 e^{-3x} \sin 2x$,   (c) $(D^2 + D - 2)y = x \cos x$,

(d) $(3D^2 + 8)y = -4 x \sin 2x$.

5.4   Homogeneous linear equations

A linear differential equation of the form

$$(a_n x^n D^n + a_{n-1} x^{n-1} D^{n-1} + \cdots + a_1 xD + a_0)y = f(x) \qquad (20)$$

is said to be *homogeneous*. The method of solution is to reduce this to a linear equation with constant coefficients by means of the substitution

$x = e^t$.

The function y can now be regarded as a function of the new variable t, and we denote by Δ^n the operator whose effect is to give the nth derivative of a function with respect to t.

PROPOSITION 5.11

$$x^n D^n = \Delta(\Delta - 1)\ldots(\Delta - n + 1) \quad for\ n = 1, 2, 3, \ldots.$$

Proof

The proof is by induction on n. Since $x = e^t$, then $dx/dt = x$ and

$$\Delta y = \frac{dy}{dt} = \frac{dy}{dx}\frac{dx}{dt} = x(Dy),$$

and the formula is valid for $n = 1$. Assume now the validity for $n = k$, i.e.

$$x^k D^k = \Delta(\Delta - 1) \ldots (\Delta - k + 1).$$

Then

$$\begin{aligned}
\Delta(\Delta(\Delta - 1) &\ldots (\Delta - k + 1)) \\
&= x(D(x^k D^k)) \\
&= x(x^k D^{k+1} + kx^{k-1}D^k) \\
&= x^{k+1}D^{k+1} + kx^k D^k \\
&= x^{k+1}D^{k+1} + k\Delta(\Delta - 1) \ldots (\Delta - k + 1).
\end{aligned}$$

Therefore

$$x^{k+1}D^{k+1} = \Delta(\Delta - 1) \ldots (\Delta - k + 1)(\Delta - k)$$

and the formula is valid for $n = k + 1$. By induction therefore it is valid for all positive integers n.

Example 14   Solve $(x^3 D^3 + 3x^2 D^2 + xD + 8)y = 65 \cos(\log x)$.

Solution   Put $x = e^t$. The equation then reduces to

$$(\Delta(\Delta - 1)(\Delta - 2) + 3\Delta(\Delta - 1) + \Delta + 8)y = 65 \cos t$$

i.e.

$$(\Delta^3 + 8)y = 65 \cos t.$$

The auxiliary equation has roots -2, $1 + i\sqrt{3}$, $1 - i\sqrt{3}$ and so the complementary function is $A\,e^{-2t} + e^t(B\cos(t\sqrt{3}) + C\sin(t\sqrt{3}))$ or

$$Ax^{-2} + x(B\cos(\sqrt{3}\log x) + C\sin(\sqrt{3}\log x)).$$

The particular integral is

$$\frac{1}{\Delta^3 + 8}\,65\cos t = \frac{1}{8 - \Delta}\,65\cos t$$

$$= (8 + \Delta)\left(\frac{1}{64 - \Delta^2}\,65\cos t\right)$$

$$= (8 + \Delta)\cos t = 8\cos t - \sin t.$$

In terms of the variable x, this is

$$8 \cos (\log x) - \sin (\log x).$$

EXERCISE (xii)
Solve

(a) $(x^3D^3 + 3x^2D^2 + xD)y = 32x^2$,   (b) $(D^2 + 2x^{-1}D)y = -x^{-3}$,
(c) $(xD^2 - D + 5x^{-1})y = \cos (\log x^2)$,
(d) $(x^2D^2 - xD - 3)y = x^3 \log x$.

CHAPTER 6

Functions of Several Variables

This chapter deals with the extension to the case of functions of several variables of certain items already dealt with for functions of one variable. Specifically, we will obtain the extension of Taylor's theorem and considerations of maxima and minima, and the solutions to standard differential equations, notably Laplace's equation and the wave equation. We will not deal here with transformations and Jacobians, nor with multiple, line or surface integrals. We envisage these important topics to be covered in a different part of the students' course.

6.1 Introduction

For simplicity of exposition we shall limit ourselves to real functions $f(x, y)$ of two independent real variables x and y. The notions of *limit* and *continuity* are quite similar to the case of one variable. The only problem is to extend the concept of *distance* (*the distance of the point x from the point α is $|x - \alpha|$*) to the case of two variables (*the distance of the point (x, y) from the point (α, β) is $|(x, y) - (\alpha, \beta)|$*). In other words, to define $|(x, y) - (\alpha, \beta)|$. Reminiscent of complex numbers, we define

$$|(x, y) - (\alpha, \beta)| = \sqrt{\{(x - \alpha)^2 + (y - \beta)^2\}}. \tag{1}$$

The definitions of Chapter 3 are then easily reformulated.

The function $f(x, y)$ is said to have limit l as (x, y) tends to (α, β) if, given any $\epsilon > 0$ there exists a δ (which depends on the choice of ϵ) such that $|f(x, y) - l| < \epsilon$ for all (x, y) satisfying

$$0 < |(x, y) - (\alpha, \beta)| < \delta.$$

The condition $0 < |(x, y) - (\alpha, \beta)| < \delta$ merely states that, if we regard the pairs (x, y), (α, β) as cartesian coordinates of points in a plane, then (x, y) lies in a circle (but not at the centre), centre (α, β) with radius δ.

As before, we write $f(x, y) \to l$ as $(x, y) \to (\alpha, \beta)$

If $f(x, y)$ is defined at (α, β) and if $f(x, y) \to f(\alpha, \beta)$ as $(x, y) \to (\alpha, \beta)$, then we say that f is *continuous* at (α, β). The analogues of the rules governing limits hold as for functions of one variable. Thus we have (cf. Proposition 3.4) the sum, product and quotient (where it is defined) of two continuous functions are continuous.

EXERCISE (i)
Suppose that

$$f(x,y) = \begin{cases} \dfrac{xy(x^2 - y^2)}{x^2 + y^2} & (x,y) \neq (0,0) \\ 0 & (x,y) = (0,0), \end{cases}$$

and

$$g(x,y) = \begin{cases} \dfrac{xy}{x^2 + y^2}, & (x,y) \neq (0,0) \\ 0, & (x,y) = (0,0). \end{cases}$$

Prove that f is continuous at $(0,0)$ but g is not continuous there. Show also that

$$h(x,y) = \frac{x^2 - y^2}{x^2 + y^2}, \qquad (x,y) \neq (0,0)$$

is not continuous at $(0,0)$ no matter how $h(0,0)$ is defined.

6.2   Partial derivatives

If the variable y is kept constant at the value $y = \beta$, then the function $f(x, \beta)$ may be regarded as a function of the single variable x. The derivative of this function at the point $x = \alpha$, if it exists, is then denoted by

$$\left. \frac{\partial f}{\partial x} \right|_{(\alpha, \beta)} \quad \text{or} \quad f_x(\alpha, \beta),$$

i.e.,

$$f_x(\alpha, \beta) = \lim_{h \to 0} \frac{f(\alpha + h, \beta) - f(\alpha, \beta)}{h}.$$

The number $f_x(\alpha, \beta)$ is called the *partial derivative of $f(x, y)$ with respect to x at the point* (α, β). If $f_x(\alpha, \beta)$ exists for some range of values for (α, β) then the function whose value at each point of the range is the partial derivative with respect to x of $f(x, y)$ at that point is denoted by $f_x(x, y)$ or $\partial f/\partial x$.

Similarly, if x is kept constant at the value $x = \alpha$, we can define

$$f_y(\alpha, \beta) = \lim_{k \to 0} \frac{f(\alpha, \beta + k) - f(\alpha, \beta)}{k}$$

and obtain the function $f_y(x, y)$ or $\partial f/\partial y$.

EXERCISE (ii)

Show that $f(x, y)$ and $g(x, y)$ of Exercise (i) have both partial derivatives at $(0, 0)$. Find the partial derivatives of these functions for $(x, y) \neq (0, 0)$ and for $(x, y) = (0, 0)$.

The functions $f_x(x, y)$ and $f_y(x, y)$ may themselves have partial derivatives. These are known as *second partial derivatives* of $f(x, y)$, and are denoted by

$$f_{xx} = \frac{\partial^2 f}{\partial x^2} = \frac{\partial}{\partial x}\left(\frac{\partial f}{\partial x}\right), \qquad f_{xy} = \frac{\partial^2 f}{\partial x\, \partial y} = \frac{\partial}{\partial x}\left(\frac{\partial f}{\partial y}\right),$$

$$f_{yx} = \frac{\partial^2 f}{\partial y\, \partial x} = \frac{\partial}{\partial y}\left(\frac{\partial f}{\partial x}\right), \qquad f_{yy} = \frac{\partial^2 f}{\partial y^2} = \frac{\partial}{\partial y}\left(\frac{\partial f}{\partial y}\right).$$

Even higher partial derivatives may exist. Note that we cannot assume in general that $f_{yx} = f_{xy}$. (It can be shown however that equality holds when both sides are continuous at the point in question, cf. Proposition 6.4.)

Example 1 Find $f_{xy}(0, 0)$ and $f_{yx}(0, 0)$ for the function $f(x, y)$ of Exercise (i).

Solution To obtain $f_x(x, y)$ for $(x, y) \neq (0, 0)$ we treat y as a constant and differentiate with respect to x; similarly for $f_y(x, y)$:

$$f_x(x, y) = \frac{y(x^4 + 4x^2 y^2 - y^4)}{(x^2 + y^2)^2} \quad \text{for } (x, y) \neq (0, 0),$$

$$f_y(x, y) = \frac{x(x^4 - 4x^2 y^2 - y^4)}{(x^2 + y^2)^2} \quad \text{for } (x, y) \neq (0, 0).$$

From the definition we obtain

$$f_x(0, 0) = 0, \quad f_y(0, 0) = 0.$$

Therefore

$$\frac{f_x(0, k) - f_x(0, 0)}{k} = -1,$$

$$\frac{f_y(h, 0) - f_y(0, 0)}{h} = 1,$$

and so $f_{xy}(0, 0) = 1$, $\quad f_{yx}(0, 0) = -1$.

EXERCISE (iii)

Find the second partial derivatives of the functions

(a) $\cos(x + y^2) + e^{xy}$, (b) $\tan^{-1}\dfrac{y}{x}$.

EXERCISE (iv)

(a) Show that $f(x, y) = \tan(x + y)$ satisfies

$$\frac{\partial f}{\partial x} - \frac{\partial f}{\partial y} = 0.$$

(b) Show that $f(x, t) = 3 \sin(x + ct) + 7 \cos(x - ct)$ satisfies

$$\frac{\partial^2 f}{\partial x^2} = \frac{1}{c^2} \frac{\partial^2 f}{\partial t^2}.$$

(c) Show that $f(x, y) = x^4 + x^3 y + 3x^2 y^2 + 27y^4$ satisfies

$$x \frac{\partial f}{\partial x} + y \frac{\partial f}{\partial y} = 4f.$$

(d) Show that $f(x, y) = e^{3y}(\sin 3x + \cos 3x) + e^{5x} \sin 5y$ satisfies

$$\frac{\partial^2 f}{\partial x^2} + \frac{\partial^2 f}{\partial y^2} = 0.$$

6.3   Differentiability

If we reexamine the definition of differentiability of a function $f(x)$ of one variable x at $x = \alpha$ we see that it is equivalent to requiring the existence of some real number A (which will depend on α and on f) such that

$$f(x) - f(\alpha) = A(x - \alpha) + \rho(x)|x - \alpha|$$

where $\rho(x) \to 0$ as $x \to \alpha$. If such an A exists then we denote it by $f'(\alpha)$.

A function $f(x, y)$ of two variables x and y is said to be *differentiable* at (α, β) is there exist real numbers A, B (which depend on α, β and on f) such that (see (1))

$$f(x, y) - f(\alpha, \beta) = A(x - \alpha) + B(y - \beta) + \rho(x, y)|(x, y) - (\alpha, \beta)|$$

$$(2)$$

where $\rho(x, y) \to 0$ as $(x, y) \to (\alpha, \beta)$. When $f(x, y)$ is differentiable then it is easy to see what A and B must be. If we put $y = \beta$, then

$$f(x, \beta) - f(\alpha, \beta) = A(x - \alpha) + \rho(x, \beta)|x - \alpha|$$

where $\rho(x, \beta) \to 0$ as $x \to \alpha$, and so $A = f_x(\alpha, \beta)$. Similarly, $B = f_y(\alpha, \beta)$. Therefore differentiability implies the existence of the first partial derivatives. The reader should be warned that the converse is not true. (The existence of continuous first partial derivatives however is sufficient to imply differentiability.)

In graphical terms the function $f(x, y)$ is differentiable at (α, β) if the surface $z = f(x, y)$ has a tangent plane at the point $(\alpha, \beta, f(\alpha, \beta))$. This tangent plane has the equation

$$z = f(\alpha, \beta) + f_x(\alpha, \beta)(x - \alpha) + f_y(\alpha, \beta)(y - \beta).$$

The linear function $f_x(\alpha, \beta)(x - \alpha) + f_y(\alpha, \beta)(y - \beta)$ is called the *derivative* of $f(x, y)$ at (α, β).

6.4   The chain rule

Suppose that $f(x, y)$ is a differentiable function of x and y. If x and y are represented as the cartesian coordinates of a point in the plane then

$$x = r \cos \theta, \qquad y = r \sin \theta$$

where r, θ are the polar coordinates of that point. The function $f(x, y)$ therefore may be regarded as a function of r and θ, and we have $f(x, y) = F(r, \theta)$ for a suitable F. The chain rule gives the relation between the partial derivatives of f and those of F. More generally

PROPOSITION 6.1   **(The chain rule)**
Suppose that $f(x, y)$ is a differentiable function of x and y at (x_0, y_0) and both $x = x(u, v)$ and $y = y(u, v)$ are themselves differentiable functions of the variables u, v at (u_0, v_0), such that

$$x_0 = x(u_0, v_0), \qquad y_0 = y(u_0, v_0).$$

Then $F(u, v) = f(x(u, v), y(u, v))$ is a differentiable function of u, v at (u_0, v_0) and

$$F_u(u_0, v_0) = f_x(x_0, y_0)x_u(u_0, v_0) + f_y(x_0, y_0) y_u(u_0, v_0), \qquad (3)$$
$$F_v(u_0, v_0) = f_x(x_0, y_0) x_v(u_0, v_0) + f_y(x_0, y_0) y_v(u_0, v_0). \qquad (4)$$

(The last two equalities are often stated in the form

$$\frac{\partial F}{\partial u} = \frac{\partial f}{\partial x} \frac{\partial x}{\partial u} + \frac{\partial f}{\partial y} \frac{\partial y}{\partial u},$$

$$\frac{\partial F}{\partial v} = \frac{\partial f}{\partial x} \frac{\partial x}{\partial v} + \frac{\partial f}{\partial y} \frac{\partial y}{\partial v},$$

or even, by abuse of notation, in the form

$$\frac{\partial f}{\partial u} = \frac{\partial f}{\partial x} \frac{\partial x}{\partial u} + \frac{\partial f}{\partial y} \frac{\partial y}{\partial u}$$

$$\frac{\partial f}{\partial v} = \frac{\partial f}{\partial x} \frac{\partial x}{\partial v} + \frac{\partial f}{\partial y} \frac{\partial y}{\partial v}.\Big)$$

Proof
Let $\epsilon_1 = |(u, v) - (u_0, v_0)|$ and $\epsilon = |(x, y) - (x_0, y_0)|$, and let C, D denote the derivatives at (u_0, v_0) of $x(u, v)$ and $y(u, v)$ respectively. Then from the differentiability of $x(u, v), y(u, v)$,

$$x(u, v) - x(u_0, v_0) - C = \rho_1 \epsilon_1,$$
$$y(u, v) - y(u_0, v_0) - D = \rho_2 \epsilon_1$$

where $\rho_1 \to 0, \rho_2 \to 0$ as $\epsilon_1 \to 0$. If we put

$$M = \max \{x_u, x_v, y_u, y_v \text{ at } (u_0, v_0), \text{ and } f_x, f_y \text{ at } (x_0, y_0)\},$$

then

$$|C| \leqslant |x_u(u_0, v_0)(u - u_0)| + |x_v(u_0, v_0)(v - v_0)| \leqslant 2M \epsilon_1$$

and

$$|D| \leqslant |y_u(u_0, v_0)(u - u_0)| + |y_v(u_0, v_0)(v - v_0)| \leqslant 2M \epsilon_1.$$

Therefore

$$\epsilon < |C + \rho_1 \epsilon_1| + |D + \rho_2 \epsilon_1| \leqslant (4M + |\rho_1| + |\rho_2|) \epsilon_1.$$

Let A, B denote the right-hand sides of (3), (4) respectively. Consider

$$F(u, v) - F(u_0, v_0) - A(u - u_0) - B(v - v_0). \tag{5}$$

This is equal to

$$f(x, y) - f(x_0, y_0) - f_x(x_0, y_0)C - f_y(x_0, y_0)D.$$

Since $f(x, y)$ is differentiable then this can be written

$$f_x(x_0, y_0)\{x(u, v) - x(u_0, v_0) - C\}$$
$$+ f_y(x_0, y_0)\{y(u, v) - y(u_0, v_0) - D\} + \rho \epsilon$$

where $\rho \to 0$ as $\epsilon \to 0$, i.e. (5) is equal to

$$\left(f_x(x_0, y_0)\rho_1 + f_y(x_0, y_0)\rho_2 + \rho \frac{\epsilon}{\epsilon_1}\right) \epsilon_1. \tag{6}$$

Since $\epsilon_1 \to 0$ implies $\rho_1 \to 0, \rho_2 \to 0$ and $\epsilon \to 0$ (and hence $\rho \to 0$), then

$$\frac{F(u, v) - F(u_0, v_0) - A(u - u_0) - B(v - v_0)}{\epsilon_1} \to 0 \quad \text{as } \epsilon_1 \to 0.$$

Therefore $F(u, v)$ is differentiable at (u_0, v_0), and A, B are the first partial derivatives.

EXAMPLE 2   If $x = r \cos \theta$ and $y = r \sin \theta$ and if $f(x, y)$ is a differentiable function of x and y, show that

$$\frac{\partial f}{\partial x} = \cos \theta \, \frac{\partial f}{\partial r} - \frac{\sin \theta}{r} \frac{\partial f}{\partial \theta},$$

$$\frac{\partial f}{\partial y} = \sin \theta \, \frac{\partial f}{\partial r} + \frac{\cos \theta}{r} \frac{\partial f}{\partial \theta}.$$

Hence find the polar form of Laplace's operator

$$\frac{\partial^2 f}{\partial x^2} + \frac{\partial^2 f}{\partial y^2}.$$

Solution   Since $r^2 = x^2 + y^2$ and $\tan \theta = y/x$, then $2r \, (\partial r/\partial x) = 2x$, $2r \, (\partial r/\partial y) = 2y$ and $\sec^2 \theta \, (\partial \theta/\partial x) = -y/x^2$, $\sec^2 \theta \, (\partial \theta/\partial y) = 1/x$. Therefore

$$\frac{\partial r}{\partial x} = \frac{x}{r} = \cos \theta, \qquad \frac{\partial r}{\partial y} = \frac{y}{r} = \sin \theta,$$

$$\frac{\partial \theta}{\partial x} = \frac{-y \cos^2 \theta}{x^2} = \frac{-y}{r^2} = \frac{-\sin \theta}{r}, \qquad \frac{\partial \theta}{\partial y} = \frac{\cos^2 \theta}{x} = \frac{\cos \theta}{r}.$$

By the chain rule,

$$\frac{\partial f}{\partial x} = \frac{\partial f}{\partial r} \frac{\partial r}{\partial x} + \frac{\partial f}{\partial \theta} \frac{\partial \theta}{\partial x} = \cos \theta \, \frac{\partial f}{\partial r} - \frac{\sin \theta}{r} \frac{\partial f}{\partial \theta},$$

$$\frac{\partial f}{\partial y} = \frac{\partial f}{\partial r} \frac{\partial r}{\partial y} + \frac{\partial f}{\partial \theta} \frac{\partial \theta}{\partial y} = \sin \theta \, \frac{\partial f}{\partial r} + \frac{\cos \theta}{r} \frac{\partial f}{\partial \theta}.$$

Now

$$\frac{\partial^2 f}{\partial x^2} = \frac{\partial}{\partial x} \left(\frac{\partial f}{\partial x} \right)$$

$$= \cos \theta \, \frac{\partial}{\partial r} \left(\cos \theta \, \frac{\partial f}{\partial r} - \frac{\sin \theta}{r} \frac{\partial f}{\partial \theta} \right) - \frac{\sin \theta}{r} \frac{\partial}{\partial \theta} \left(\cos \theta \, \frac{\partial f}{\partial r} - \frac{\sin \theta}{r} \frac{\partial f}{\partial \theta} \right)$$

$$= \cos \theta \left(\cos \theta \, \frac{\partial^2 f}{\partial r^2} - \frac{\sin \theta}{r} \frac{\partial^2 f}{\partial r \, \partial \theta} + \frac{\sin \theta}{r^2} \frac{\partial f}{\partial \theta} \right)$$

$$- \frac{\sin \theta}{r} \left(\cos \theta \, \frac{\partial^2 f}{\partial \theta \, \partial r} - \sin \theta \, \frac{\partial f}{\partial r} - \frac{\sin \theta}{r} \frac{\partial^2 f}{\partial \theta^2} - \frac{\cos \theta}{r} \frac{\partial f}{\partial \theta} \right)$$

$$= \cos^2 \theta \, \frac{\partial^2 f}{\partial r^2} - \frac{2 \sin \theta \cos \theta}{r} \frac{\partial^2 f}{\partial r \, \partial \theta} + \frac{2 \sin \theta \cos \theta}{r^2} \frac{\partial f}{\partial \theta}$$

$$+ \frac{\sin^2 \theta}{r} \frac{\partial f}{\partial r} + \frac{\sin^2 \theta}{r^2} \frac{\partial^2 f}{\partial \theta^2}$$

since $\partial^2 f/\partial r\partial\theta = \partial^2 f/\partial\theta\,\partial r$ (we assume here that both $\partial^2 f/\partial r\partial\theta$ and $\partial^2 f/\partial\theta\partial r$ are continuous—c.f. Proposition 6.4). Similarly (this proof should be written out at least once by the student)

$$\frac{\partial^2 f}{\partial y^2} = \sin^2\theta\,\frac{\partial^2 f}{\partial r^2} + \frac{2\sin\theta\cos\theta}{r}\,\frac{\partial^2 f}{\partial r\,\partial\theta} - \frac{2\sin\theta\cos\theta}{r^2}\,\frac{\partial f}{\partial\theta}$$

$$+ \frac{\cos^2\theta}{r}\,\frac{\partial f}{\partial r} + \frac{\cos^2\theta}{r^2}\,\frac{\partial^2 f}{\partial\theta^2}\,.$$

Therefore

$$\frac{\partial^2 f}{\partial x^2} + \frac{\partial^2 f}{\partial y^2} = \frac{\partial^2 f}{\partial r^2} + \frac{1}{r}\,\frac{\partial f}{\partial r} + \frac{1}{r^2}\,\frac{\partial^2 f}{\partial\theta^2}\,.$$

EXERCISE (v)

(a) If $x = r\cos\theta$, $y = r\sin\theta$, calculate the products

$$\frac{\partial r}{\partial x}\,\frac{\partial x}{\partial r}\,,\qquad \frac{\partial r}{\partial y}\,\frac{\partial y}{\partial r}\,,\qquad \frac{\partial\theta}{\partial x}\,\frac{\partial x}{\partial\theta}\,,\qquad \frac{\partial\theta}{\partial y}\,\frac{\partial y}{\partial\theta}$$

and be warned that in no case is the product equal to 1.

(b) If $f(x,y)$ is a function of x and y and if $u = e^x$, $v = e^y$, show that $\partial^2 f/\partial x\partial y = uv\,(\partial^2 f/\partial u\partial v)$.

EXERCISE (vi)

Suppose $f(x,y)$ is a function of x and y where $x = u^2 - v^2$, $y = 2uv$. Show that

$$\left(\frac{\partial f}{\partial x}\right)^2 + \left(\frac{\partial f}{\partial y}\right)^2 = 4(x^2 + y^2)\left\{\left(\frac{\partial f}{\partial u}\right)^2 + \left(\frac{\partial f}{\partial v}\right)^2\right\}.$$

EXERCISE (vii)

Suppose $x = e^u\cos v$, $y = e^u\sin v$. Show that any function $f(x,y)$ which

(a) satisfies $\left(\dfrac{\partial f}{\partial x}\right)^2 + \left(\dfrac{\partial f}{\partial y}\right)^2 = 0$ also satisfies $\left(\dfrac{\partial f}{\partial u}\right)^2 + \left(\dfrac{\partial f}{\partial v}\right)^2 = 0$;

(b) satisfies $\dfrac{\partial^2 f}{\partial x^2} + \dfrac{\partial^2 f}{\partial y^2} = 0$ also satisfies $\dfrac{\partial^2 f}{\partial u^2} + \dfrac{\partial^2 f}{\partial v^2} = 0$.

Special cases of the chain rule are obtained in the following circumstances.

PROPOSITION 6.2
If $f(u)$ is a differentiable function of one variable u, where $u = u(x, y)$ is a differentiable function of x and y, then

$$\frac{\partial f}{\partial x} = \frac{df}{du} \frac{\partial u}{\partial x}, \qquad \frac{\partial f}{\partial y} = \frac{df}{du} \frac{\partial u}{\partial y}.$$

PROPOSITION 6.3 **(Theorem of total differentiation)**
If $f(x, y)$ is differentiable at all points of a curve $x = x(t)$, $y = y(t)$, where $x(t), y(t)$ are differentiable functions, then $f(x, y)$ is a differentiable function of t and

$$\frac{df}{dt} = \frac{\partial f}{\partial x} \frac{dx}{dt} + \frac{\partial f}{\partial y} \frac{dy}{dt}.$$

EXERCISE (viii)
If φ and ψ are arbitrary differentiable functions of one variable and if $f(x, t) = \varphi(x + ct) + \psi(x - ct)$ where c is a constant, show that f is a solution of the *wave equation*

$$\frac{\partial^2 f}{\partial x^2} = \frac{1}{c^2} \frac{\partial^2 f}{\partial t^2}.$$

EXERCISE (ix)
Suppose that $f(tx, ty) = t^N f(x, y)$ for some fixed positive integer N whenever $t > 0$ (f is then said to be *homogeneous* of degree N). If f is differentiable at (α, β), show that $xf_x(\alpha, \beta) + yf_y(\alpha, \beta) = Nf(\alpha, \beta)$ (Euler's theorem).

Example 3   Find general solutions of the wave equation.

Solution   Put $u = x + ct, v = x - ct$. By the chain rule

$$\frac{\partial f}{\partial x} = \frac{\partial f}{\partial u} \frac{\partial u}{\partial x} + \frac{\partial f}{\partial v} \frac{\partial v}{\partial x} = \frac{\partial f}{\partial u} + \frac{\partial f}{\partial v},$$

$$\frac{\partial f}{\partial t} = \frac{\partial f}{\partial u} \frac{\partial u}{\partial t} + \frac{\partial f}{\partial v} \frac{\partial v}{\partial t} = c \left(\frac{\partial f}{\partial u} - \frac{\partial f}{\partial v} \right).$$

Therefore

$$\frac{\partial^2 f}{\partial x^2} = \left(\frac{\partial}{\partial u} + \frac{\partial}{\partial v} \right) \left(\frac{\partial f}{\partial u} + \frac{\partial f}{\partial v} \right) = \frac{\partial^2 f}{\partial u^2} + 2 \frac{\partial^2 f}{\partial u \, \partial v} + \frac{\partial^2 f}{\partial v^2},$$

$$\frac{\partial^2 f}{\partial t^2} = c^2 \left(\frac{\partial}{\partial u} - \frac{\partial}{\partial v} \right) \left(\frac{\partial f}{\partial u} - \frac{\partial f}{\partial v} \right) = c^2 \left(\frac{\partial^2 f}{\partial u^2} - 2 \frac{\partial^2 f}{\partial u \, \partial v} + \frac{\partial^2 f}{\partial v^2} \right).$$

Since

$$\frac{\partial^2 f}{\partial x^2} - \frac{1}{c^2}\frac{\partial^2 f}{\partial t^2} = 4\frac{\partial^2 f}{\partial u\,\partial v}$$

then we must solve

$$\frac{\partial^2 f}{\partial u\,\partial v} = 0.$$

Integrating, this gives $\partial f/\partial v = g(v)$ where g is an arbitrary function. Therefore integrating with respect to v gives

$$f = \int g(v)\,dv + \varphi(u)$$

$$= \psi(v) + \varphi(u)$$

where $\psi(v) = \int g(v)\,dv$. General solutions of the wave equation therefore are

$$f(x, t) = \varphi(x + ct) + \psi(x - ct).$$

6.5 Taylor's theorem

We can extend Taylor's theorem to functions of several variables. First of all we state and prove the First Mean Value Theorem, as this serves also to illustrate the general method.

PROPOSITION 6.3   (First Mean Value Theorem)
Suppose that $f(x, y)$ is differentiable at all points of the line segment PQ joining $P = (\alpha, \beta)$ to $Q = (\alpha + h, \beta + k)$. Then

$$f(\alpha + h, \beta + k) = f(\alpha, \beta) + R_1$$

where $R_1 = hf_x(\alpha + \theta h, \beta + \theta k) + kf_y(\alpha + \theta h, \beta + \theta k)$ and $0 < \theta < 1$.

Proof
Consider the function $F(t) = f(\alpha + th, \beta + tk)$. By the theorem of total differentiation $F'(t) = hf_x(\alpha + th, \beta + tk) + kf_y(\alpha + th, \beta + tk)$. From the First Mean Value Theorem for functions of one variable

$$F(1) = F(0) + F'(\theta)$$

where $0 < \theta < 1$. Since $F(1) = f(\alpha + h, \beta + k)$, $F(0) = f(\alpha, \beta)$ and $F'(\theta) = R_1$, this yields the formula.
    An application of the MVT is

PROPOSITION 6.4
If f_{xy} and f_{yx} are both continuous then they are equal.

Proof

Suppose that f_{xy} and f_{yx} are both continuous at (α, β). Put $\varphi(x) = f(x, \beta + k) - f(x, \beta)$. Then

$$\varphi(\alpha + h) - \varphi(\alpha) = h\varphi'(\alpha + \theta_1 h) = h\{f_x(\alpha + \theta_1 h, \beta + k) - \\ - f_x(\alpha + \theta_1 h, \beta)\} \\ = hkf_{yx}(\alpha + \theta_1 h, \beta + \theta_2 k)$$

where $0 < \theta_1 < 1, 0 < \theta_2 < 1$. Similarly, by taking $\psi(y) = f(\alpha + h, y) - f(\alpha, y)$ we obtain

$$\psi(\beta + k) - \psi(\beta) = khf_{xy}(\alpha + \theta_3 h, \beta + \theta_4 k)$$

where $0 < \theta_3 < 1, 0 < \theta_4 < 1$. But $\varphi(\alpha + h) - \varphi(\alpha) = \psi(\beta + k) - \psi(\beta)$, and so

$$f_{yx}(\alpha + \theta_1 h, \beta + \theta_2 k) = f_{xy}(\alpha + \theta_3 h, \beta + \theta_4 k).$$

From the continuity of these functions, the left-hand side tends to $f_{yx}(\alpha, \beta)$ and the right-hand side tends to $f_{xy}(\alpha, \beta)$ as $h \to 0, k \to 0$. Hence $f_{yx}(\alpha, \beta) = f_{xy}(\alpha, \beta)$.

This proposition extends to higher derivatives: the order in which partial derivatives are formed is irrelevant provided these derivatives are continuous. We shall not prove this.

PROPOSITION 6.5 **(Taylor's theorem for functions of two variables)**
Suppose that $f(x, y)$ has continuous partial derivatives up to the $(N + 1)$st order for all points in a disc, centre (α, β). Then for any point $(\alpha + h, \beta + k)$ in this disc

$$f(\alpha + h, \beta + k) = f(\alpha, \beta) + \left(h\frac{\partial}{\partial x} + k\frac{\partial}{\partial y}\right) f(\alpha, \beta)$$

$$+ \frac{1}{2!} \left(h\frac{\partial}{\partial x} + k\frac{\partial}{\partial y}\right)^2 f(\alpha, \beta) + \cdots$$

$$+ \frac{1}{N!} \left(h\frac{\partial}{\partial x} + k\frac{\partial}{\partial y}\right)^N f(\alpha, \beta) + R_N,$$

where $R_N = 1/(N + 1)! \, (h(\partial/\partial x) + k(\partial/\partial y))^{N+1} f(\alpha + \theta h, \beta + \theta k)$, $0 < \theta < 1$. [The notation $(h(\partial/\partial x) + k(\partial/\partial y))^n f(\alpha, \beta)$ is taken to mean that the operator is first expanded by the binomial theorem, then applied to $f(x, y)$ and the resulting function evaluated at (α, β).]

Proof

Consider the function $F(t) = f(\alpha + th, \beta + tk)$. By the theorem of total differentiation

$$F'(t) = \left(h\frac{\partial}{\partial x} + k\frac{\partial}{\partial y}\right) f(\alpha + th, \beta + tk),$$

and in general

$$F^n(t) = \left(h\frac{\partial}{\partial x} + k\frac{\partial}{\partial y} \right)^n f(\alpha + th, \beta + tk).$$

Substituting in Taylor's theorem (for one variable) yields the required result immediately.

6.6 Maxima and minima

A function $f(x, y)$ is said to have a *local maximum* at (α, β) if (α, β) is the centre of some disc throughout which $f(x, y) \leqslant f(\alpha, \beta)$, i.e. if there exists $\delta > 0$ such that the inequality holds whenever $0 < |(x, y) - (\alpha, \beta)| < \delta$. If the inequality is reserved then we speak of a *local minimum.* We shall use the terms *maximum* and *minimum* for brevity, and say that f has an *extreme value* at these points.

PROPOSITION 6.6
If $f(x, y)$ has an extreme value at (α, β) and if the first partial derivatives exist at (α, β) then $f_x(\alpha, \beta) = 0 = f_y(\alpha, \beta)$.

Proof
Consider the case of a maximum. The function of one variable $F(x) = f(x, \beta)$ clearly has a maximum at $x = \alpha$. By Proposition 4.8 $F'(\alpha) = 0$, and since $F'(x) = f_x(x, \beta)$, then $f_x(\alpha, \beta) = 0$. Similarly, $G(y) = f(\alpha, y)$ has a maximum at $y = \beta$, and so $f_y(\alpha, \beta) = G'(\beta) = 0$. The case of a minimum is proved similarly.

As we have noted for functions of one variable (§4.4) the vanishing of the first derivative at a point is not sufficient for a function to have a maximum or a minimum there. The point (α, β) is called a *stationary point* if $f_x(\alpha, \beta) = 0 = f_y(\alpha, \beta)$.

Example 4   Examine the stationary points of $f(x, y) = y^2 - x^2$.

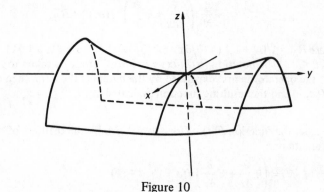

Figure 10

Solution Since $f_x(x, y) = -2x, f_y(x, y) = 2y$ the only stationary point occurs at the origin $(0, 0)$. The value of the function at the origin is 0. Since any disc centre the origin contains points $(0, y)$(where $f(0, y) > 0$) and points $(x, 0)$ (where $f(x, 0) < 0$) then f cannot have a maximum or a minimum at $(0, 0)$. The graph of this function is illustrated in Fig. 10.

PROPOSITION 6.7

Suppose that $f(x, y)$ has a stationary point at (α, β) and its second order partial derivatives are continuous there. If

$$\{f_{xy}(\alpha, \beta)\}^2 < f_{xx}(\alpha, \beta) f_{yy}(\alpha, \beta),$$

then f has a maximum at (α, β) if $f_{xx}(\alpha, \beta) < 0$, and a minimum at (α, β) if $f_{xx}(\alpha, \beta) > 0$. However, if

$$\{f_{xy}(\alpha, \beta)\}^2 > f_{xx}(\alpha, \beta) f_{yy}(\alpha, \beta),$$

then f has neither a maximum nor a minimum at (α, β).

Proof

Put $A = f_{xx}(\alpha, \beta), B = f_{yy}(\alpha, \beta)$ and $H = f_{xy}(\alpha, \beta)$. Since $f_x(\alpha, \beta) = 0 = f_y(\alpha, \beta)$, then Taylor's theorem (with $N = 1$) yields

$$f(\alpha + h, \beta + k) - f(\alpha, \beta) = \tfrac{1}{2} \{h^2 f_{xx}(\alpha + \theta h, \beta + \theta k) + 2hk f_{xy}(\alpha + \theta h,$$
$$\beta + \theta k) + k^2 f_{yy}(\alpha + \theta h, \beta + \theta k)\}$$

where $0 < \theta < 1$. From the continuity of the second partial derivatives, we deduce that, for small values of h and k, the right hand side of this equation takes the same sign as

$$h^2 A + 2hkH + k^2 B. \tag{7}$$

Suppose that $AB - H^2 > 0$. If $A > 0$, we put $D^2 = A$ and note that the expression (7) may be written in the form

$$\left(Dh + \frac{Hk}{D}\right)^2 + k^2 \left(\frac{AB - H^2}{D^2}\right) \tag{8}$$

which is positive for all values of h, k, and so f has a minimum at (α, β). If $A < 0$, we put $D^2 = -A$ and (7) becomes

$$-\left(Dh - \frac{Hk}{D}\right)^2 - k^2 \left(\frac{AB - H^2}{D^2}\right) \tag{9}$$

which is negative for all values of h, k, and so f has a maximum at (α, β). Suppose on the other hand that $AB - H^2 < 0$. Then the expression (8) takes positive values when $k = 0$ and negative values when $h = -Hk/D^2$. Similarly the expression (9) is positive when $h = Hk/D^2$ and negative when $k = 0$. Therefore f has neither a maximum nor a minimum at (α, β). Note: this argument could be greatly simplified by the use of quadratic forms.

EXERCISE (x)

Find the stationary points of the following functions and investigate their nature:

(a) $(3 + x - y)xy$, (b) $x^2 + y^2 + xy + x + y$,
(c) $x^2 + \frac{1}{4}y^2 - xy + x + y$, (d) $e^{-x^2}(xy + 2x - y)$.

EXERCISE (xi)

Find the point on the plane $x + 3y - z + 8 = 0$ which is nearest to the point $(1, 1, 1)$.

Sufficient conditions may similarly be obtained for functions $f(x, y, z)$ of three variables x, y and z. We merely state the result here. Let

$$A = f_{xx}, \quad B = f_{yy}, \quad C = f_{zz}, \quad F = f_{yz}, \quad G = f_{zx},$$
$$H = f_{xy}$$

evaluated at (α, β, γ). Suppose also that $f_x(\alpha, \beta, \gamma) = f_y(\alpha, \beta, \gamma) = f_z(\alpha, \beta, \gamma) = 0$. Then (in determinantal notation) f has a minimum at (α, β, γ) if

$$A > 0, \quad \begin{vmatrix} A & H \\ H & B \end{vmatrix} > 0, \quad \begin{vmatrix} A & H & G \\ H & B & F \\ G & F & C \end{vmatrix} > 0,$$

and a maximum at (α, β, γ) if

$$A < 0, \quad \begin{vmatrix} A & H \\ H & B \end{vmatrix} > 0, \quad \begin{vmatrix} A & H & G \\ H & B & F \\ G & F & C \end{vmatrix} < 0.$$

6.7 Lagrange multipliers

Sometimes it is necessary to find the extreme values of a function $f(x, y)$ when (x, y) is restricted to lie on a curve Γ with equation $\varphi(x, y) = 0$. If the equation of the curve can be expressed in the form $y = \psi(x)$ then our problem can be solved by the methods of §4.4 applied to $f(x, \psi(x))$. In many cases however this method is impracticable, and instead we use *Lagrange's undetermined multipliers*.

Suppose that $x = x(t), y = y(t)$ is a parametric representation of Γ. Then $f(x, y)$ becomes a function of t for points on the curve. It follows then that if f becomes stationary at the point (α, β) on Γ then $df/dt = 0$ there, i.e.

$$\frac{\partial f}{\partial x}\frac{dx}{dt} + \frac{\partial f}{\partial y}\frac{dy}{dt} = 0. \tag{10}$$

Since $\varphi(x(t), y(t)) = 0$, then $d\varphi/dt = 0$ and so

$$\frac{\partial\varphi}{\partial x}\frac{dx}{dt} + \frac{\partial\varphi}{\partial y}\frac{dy}{dt} = 0. \tag{11}$$

If φ_x and φ_y are not both zero at (α, β) then there exists a number λ such that $f + \lambda\varphi$ satisfies the first order equations

$$f_x + \lambda\varphi_x = 0, \ f_y + \lambda\varphi_y = 0 \quad \text{at } (\alpha, \beta). \tag{12}$$

The equations (12) together with

$$\varphi(\alpha, \beta) = 0$$

are then used to determine α, β. It is often unnecessary to determine the value of λ, hence the name: *undetermined multiplier*.

Example 5   Find the stationary values of $f(x, y) = x^2 + y^2$ subject to the condition that $x^3 + y^3 - 12xy = 0$.

Solution   By the method of Lagrange's undetermined multipliers there exists λ such that the function

$$x^2 + y^2 + \lambda(x^3 + y^3 - 12xy)$$

yields the first order equations

$$2x + \lambda(3x^2 - 12y) = 0,$$
$$2y + \lambda(3y^2 - 12x) = 0.$$

Therefore

$$\frac{2x}{3x^2 - 12y} = -\lambda = \frac{2y}{3y^2 - 12x},$$

and so

$$x(3y^2 - 12x) = y(3x^2 - 12y)$$

or

$$-xy^2 + 4x^2 + yx^2 - 4y^2 = 0.$$

This may be rewritten in the form

$$(x - y)(xy + 4x + 4y) = 0.$$

Solving simultaneously

$$x - y = 0,$$

and

$$x^2 + y^3 - 12xy = 0$$

Yields $x = 0 = y$ and $x = 6 = y$. Solving simultaneously

$$xy + 4x + 4y = 0,$$

and

$$x^3 + y^3 - 12xy = 0$$

yields $x = 0 = y$ as the only real solution. Therefore the stationary values of $f(x, y)$ occur among the points $(0, 0)$ and $(6, 6)$. Since $f(x, y)$ represents (but is not *equal* to) the distance from the origin, the point $(0, 0)$ clearly yields a minimum, while $(6, 6)$ yields a local maximum of this distance (see Fig. 11).

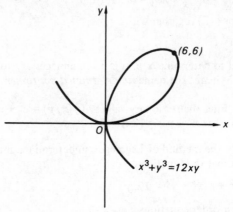

Figure 11

The method of Lagrange multipliers extends to functions of any number of variables subject to a lesser number of constraints. Example 6 illustrates the method for three variables with one constraint, while in Example 7 the three variables are subject to two constraints.

Example 6   Find the dimensions of the box of greatest volume which can be fitted into the ellipsoid $2x^2 + 4y^2 + z^2 = 12$ in such a way that each edge of the box is parallel to a coordinate axis.

Solution   If the corner of the box in the first octant has coordinates (x, y, z), then the dimensions are $2x$, $2y$ and $2z$, and the volume V is $8xyz$. Also, (x, y, z) will lie on the ellipsoid. By Lagrange's method, there exists λ such that

$$8xyz + \lambda(2x^2 + 4y^2 + z^2 - 12)$$

yields the first order equations

$$8yz + \lambda(4x) = 0,$$
$$8zx + \lambda(8y) = 0, \qquad\qquad (13)$$
$$8xy + \lambda(2z) = 0.$$

Multiplying by x, y, z respectively and adding we obtain

$$24xyz + \lambda(4x^2 + 8y^2 + 2z^2) = 0.$$

But since $2x^2 + 4y^2 + z^2 = 12$, then

$$xyz = -\lambda.$$

Substituting in (13) gives $x^2 = 2, y^2 = 1, z^2 = 4$ (since $xyz \neq 0$) and so $x = \sqrt{2}, y = 1, z = 2$.

Example 7   Find the least distance from the origin to the curve of intersection of the surface $x^2 + 2y^2 + z^2 - 2yz - 2xy = 2$ and the plane $x + y + z = 3$.

Solution   We wish to find the least value of $x^2 + y^2 + z^2$ subject to the constraints $x^2 + 2y^2 + z^2 - 2yz - 2xy - 2 = 0$ and $x + y + z - 3 = 0$. By Lagrange's method there are multipliers λ, μ such that

$$F(x, y, z) = x^2 + y^2 + z^2 + \lambda(x^2 + 2y^2 + z^2 - 2yz - 2xy - 2)$$
$$+ \mu(x + y + z - 3)$$

satisfies the first order equations $F_x = 0, F_y = 0, F_z = 0$, i.e.

$$2x + \lambda(2x - 2y) + \mu = 0, \tag{i}$$
$$2y + \lambda(4y - 2z - 2x) + \mu = 0, \tag{ii}$$
$$2z + \lambda(2z - 2y) + \mu = 0. \tag{iii}$$

We bear in mind also that

$$x^2 + 2y^2 + z^2 - 2yz - 2xy = 2, \tag{iv}$$
$$x + y + z = 3. \tag{v}$$

If we add (i), (ii) and (iii) and use (v) we obtain

$$\mu = -2.$$

Subtracting (iii) from (i) gives

$$(x - z)(1 + \lambda) = 0. \tag{vi}$$

If $\lambda = -1$, then from (i) $y = 1$, and so substituting in (iv) and (v) and solving, this gives $x = 0, z = 2$ or $x = 2, z = 0$. Therefore equations (i)–(v) are satisfied by $(0, 1, 2), (2, 1, 0), \lambda = -1$ and $\mu = -2$.
If $x = z$ (cf. (vi)) then (iv) and (v) yields

$$2x + y = 3,$$
$$x^2 + y^2 - 2xy = 1 \quad \text{or} \quad (x - y)^2 = 1.$$

These simultaneous equations yield the points $(4/3, 1/3, 4/3)$ and $(2/5, 5/3, 2/3)$, with $\lambda = -1/3, \mu = -2$.
By comparing the distances from the origin of the four points thus obtained, we see that the least distance is $\sqrt{(11/3)}$, achieved at the points $(4/3, 1/3, 4/3)$ and $(2/3, 5/3, 2/3)$.

EXERCISE (xii)
If $f(x, y, z) = 2x^2 + y^2 + 3z^2$ and the points (x, y, z) are restricted to lie in the plane $x + y + z = 11$, show that the value of f is stationary at $(3, 6, 2)$ and that f assumes its least value there.

EXERCISE (xiii)
Find the greatest value of $x + 2y + z$ subject to the restriction $x^2 + 4y^2 + 4z^2 = 1$.

EXERCISE (xiv)
Show that the shortest line from the point $(1, 3, 2)$ to the circle formed by the intersection of the sphere $x^2 + y^2 + z^2 = 194$ and the plane $x + 2y + 3z = 4$ meets the circle at $(1, 12, -7)$.

6.8 Partial differential equations

We have seen (Example 3) that the general solution of the *one-dimensional wave equation*

$$\frac{\partial^2 f}{\partial x^2} = \frac{1}{c^2} \frac{\partial^2 f}{\partial t^2} \tag{14}$$

is $f(x, t) = \varphi(x + ct) + \psi(x - ct)$ where φ and ψ are arbitrary functions. Other important equations are the (*two-dimensional*) *Laplace equation*

$$\frac{\partial^2 f}{\partial x^2} + \frac{\partial^2 f}{\partial y^2} = 0 \tag{15}$$

and the *equation of linear heat conduction and diffusion*

$$\frac{\partial^2 f}{\partial x^2} = \frac{1}{k} \frac{\partial f}{\partial t} . \tag{16}$$

We shall not seek to find general solutions, but instead will investigate solutions which take a particular form, namely one where the *variables separate*.

Example 8   Find solutions of the wave equation of the form $f(x, t) = X(x) T(t)$.

Solution   Suppose that $f(x, y) = X(x) T(t)$ is a solution of (14), where $X(x)$ is a function of x alone and $T(t)$ is a function of t alone. Substituting in (14) we obtain

$$T \frac{d^2 X}{dx^2} = \frac{1}{c^2} X \frac{d^2 T}{dt^2} ,$$

or

$$\frac{1}{X}\frac{d^2X}{dx^2} = \frac{1}{c^2T}\frac{d^2T}{dt^2}.$$

Since the left-hand side is a function of x alone, and the right-hand side is a function of t alone, and since x and t are independent variables, then there exists a *constant* h such that

$$\frac{1}{X}\frac{d^2X}{dx^2} = h = \frac{1}{c^2T}\frac{d^2T}{dt^2},$$

i.e.

$$\frac{d^2X}{dx^2} = hX, \qquad \frac{d^2T}{dt^2} = hc^2T.$$

If $h < 0$, then $h = -k^2$ say, and these equations have the solutions

$X = A\cos kx + B\sin kx,$

$T = C\cos kct + D\sin kct$

respectively. (The case $h > 0$, i.e. $h = k^2$ may be ignored as the resulting solution for T is $T = C\,e^{kct} + D\,e^{-kct}$, which is not *periodic*: an essential property of wave motion!)

It should be noted that any sum of such solutions described in Example 8 will also be a solution of the wave equation.

We may continue the exercise by imposing certain boundary conditions on our problem. For example, if one end of our string (whose motion is described by the wave equation) is fixed, say at $x = 0$, then $f(0, t) = 0$ for all t, and so $A = 0$ and

$$XT = \sin kx\,(C\cos kct + D\sin kct).$$

If the string is of length l and both ends are fixed then also $f(l, t) = 0$ for all t, and so $X(l) = 0$ or $\sin kl = 0$. Therefore $kl = n\pi$ for some integer n and

$$XT = \sin\frac{n\pi x}{l}\left(C\cos\frac{n\pi ct}{l} + D\sin\frac{n\pi ct}{l}\right).$$

In view of our remark about sums of solutions being a solution, we have

$$f(x, t) = \sum_{n=1}^{\infty}\sin\frac{n\pi x}{l}\left(C_n\cos\frac{n\pi ct}{l} + D_n\sin\frac{n\pi ct}{l}\right) \qquad (17)$$

is a solution of the wave equation, provided the coefficients C_n, D_n are chosen so that (17) converges.

Suppose that the initial shape and the initial motion of the string are given by the conditions

$$f(x, 0) = \varphi(x),$$
$$f_t(x, 0) = \psi(x)$$

respectively. This will be sufficient information for us to determine C_n, D_n. The first of our conditions implies that

$$\varphi(x) = \sum_{n=1}^{\infty} C_n \sin \frac{n\pi x}{l},$$

while the second implies that

$$\psi(x) = \sum_{n=1}^{\infty} D_n \frac{n\pi c}{l} \sin \frac{n\pi x}{l}.$$

For any positive integer m

$$\int_0^l \varphi(x) \sin \frac{m\pi x}{l} \, dx = \int_0^l \left(\sum_{n=1}^{\infty} C_n \sin \frac{n\pi x}{l} \sin \frac{m\pi x}{l} \right) dx$$

$$= \sum_{n=1}^{\infty} C_n \left(\int_0^l \sin \frac{n\pi x}{l} \sin \frac{m\pi x}{l} \, dx \right) \quad (?)$$

$$= \frac{1}{2} \sum_{n=1}^{\infty} C_n \int_0^l \left(\cos \frac{(m-n)\pi x}{l} - \cos \frac{(m+n)\pi x}{l} \right) dx$$

$$= \tfrac{1}{2} l C_m \tag{18}$$

since $\int_0^l \cos (r\pi x/l) \, dx = l$ if $r = 0$ and equals 0 if $r \neq 0$. (The step (?) has to be justified—we omit the justification here.) Therefore

$$C_n = \frac{2}{l} \int_0^l \varphi(x) \sin \frac{n\pi x}{l} \, dx,$$

and similarly,

$$\frac{n\pi c}{l} D_n = \frac{2}{l} \int_0^l \psi(x) \sin \frac{n\pi x}{l} \, dx.$$

The equality (18) can be restated in the form

LEMMA 6.8
If

$$\varphi(x) = \sum_{n=1}^{\infty} C_n \sin \frac{n\pi x}{l},$$

then

$$C_n = \frac{2}{l} \int_0^l \varphi(x) \sin \frac{n\pi x}{l} \, dx.$$

Example 9   Use the method of separation of variables to find a solution of the wave equation which represents the motion of a vibrating string of unit length, and which satisfies the boundary conditions: the string is fixed at both ends, initially at rest, and has the initial shape $f(x, 0) = \sin^3 \pi x$.

Solution   Since the string is initially at rest, then $\psi(x) = 0$ and so $D_n = 0$ for all n. Put $l = 1$, then for all x between 0 and 1

$$\sum_{n=1}^{\infty} C_n \sin n\pi x = f(x, 0) = \sin^3 \pi x = \tfrac{3}{4} \sin \pi x - \tfrac{1}{4} \sin 3\pi x.$$

Therefore $C_1 = \tfrac{3}{4}$, $C_3 = -\tfrac{1}{4}$ and $C_n = 0$ for $n \neq 1$ or 3, and so from (17)

$$f(x, t) = \tfrac{3}{4} \sin \pi x \cos \pi ct - \tfrac{1}{4} \sin 3\pi x \cos 3\pi ct.$$

Example 10   Find solutions of Laplace's equation of the form $\varphi(r) \, \psi(\theta)$, and obtain the most general form for $\varphi(r)$ when $\psi(\theta) = \cos 2\theta$.

Solution   From Example 2, Laplace's equation in polar coordinates is

$$\frac{\partial^2 f}{\partial r^2} + \frac{1}{r} \frac{\partial f}{\partial r} + \frac{1}{r^2} \frac{\partial^2 f}{\partial \theta^2} = 0.$$

Substituting $f(r, \theta) = \varphi(r)\psi(\theta)$, we obtain

$$\psi \frac{d^2 \varphi}{dr^2} + \frac{1}{r} \psi \frac{d\varphi}{dr} + \frac{1}{r^2} \varphi \frac{d^2 \psi}{d\theta^2} = 0,$$

and so

$$\frac{1}{\varphi} \left\{ r^2 \frac{d^2 \varphi}{dr^2} + r \frac{d\varphi}{dr} \right\} = -\frac{1}{\psi} \frac{d^2 \psi}{d\theta^2}.$$

The left-hand side is a function of r alone, and the right-hand side is a function of θ alone. Therefore, for some constant a,

$$r^2 \frac{d^2\varphi}{dr^2} + r\frac{d\varphi}{dr} = a\varphi,$$

$$\frac{d^2\psi}{d\theta^2} = -a\psi.$$

Since ψ must be periodic (i.e. $\psi(\theta) = \psi(\theta + 2\pi)$) then a must be positive, say $a = c^2$, and we have

$$\psi(\theta) = A \cos c\theta + B \sin c\theta,$$
$$\varphi(r) = Cr^c + Dr^{-c}.$$

If $\psi(\theta) = \cos 2\theta$, then $A = 1, B = 0$ and $c = 2$, and so

$$\varphi(r) = Cr^2 + Dr^{-2}.$$

Example 11 Use the method of separation of variables to find a solution of Laplace's equation for points inside the square $x = 0, x = 1$, $y = 0, y = 1$ which takes the value $x(1 - x)$ on the side $y = 0$ and vanishes on the other three sides.

Solution Suppose $X(x) Y(y)$ satisfies Laplace's equation (15). Then

$$Y\frac{d^2X}{dx^2} + X\frac{d^2Y}{dy^2} = 0$$

this means that, for some constant k,

$$\frac{d^2X}{dx^2} = -k^2X \quad \text{and} \quad \frac{d^2Y}{dy^2} = k^2Y,$$

or

$$\frac{d^2X}{dx^2} = k^2X \quad \text{and} \quad \frac{d^2Y}{dy^2} = -k^2Y.$$

Thus a solution of Laplace's equation is obtained by taking any linear combinations of the functions

$$e^{ky} \cos kx, \quad e^{ky} \sin kx, \quad e^{kx} \cos ky, \quad e^{kx} \sin ky,$$

for varying values of k. Since our solution $f(x, y)$ vanishes for $x = 0$, then we need take only combinations of $e^{ky} \sin kx$. Also, $f(1, y) = 0$ implies that $k = n\pi$. Therefore

$$f(x, y) = \sum_{n=1}^{\infty} \sin n\pi x (A_n e^{n\pi y} + B_n e^{-n\pi y}),$$

i.e.

$$f(x, y) = \sum_{n=1}^{\infty} \sin n\pi x (C_n \cosh n\pi y + D_n \sinh n\pi y).$$

Since $f(x, 1) = 0 = \sum_{n=1}^{\infty} \sin n\pi x (C_n \cosh n\pi + D_n \sinh n\pi)$, then

$$C_n = E_n \sinh n\pi, \qquad D_n = -E_n \cosh n\pi$$

for some constant E_n. Thus

$$f(x, y) = \sum_{n=1}^{\infty} E_n \sin n\pi x (\sinh n\pi \cosh n\pi y - \cosh n\pi \sinh n\pi y)$$

$$= \sum_{n=1}^{\infty} E_n \sin n\pi x \sinh n\pi (1 - y).$$

Finally

$$f(x, 0) = x(1 - x) = \sum_{n=1}^{\infty} E_n \sinh n\pi \sin n\pi x.$$

By Lemma 6.8,

$$E_n \sinh n\pi = 2 \int_0^1 x(1 - x) \sin n\pi x \, dx$$

$$= \frac{4}{n^3 \pi^3} (1 - (-1)^n).$$

Therefore $E_{2m} = 0, E_{2m+1} \sinh (2m + 1)\pi = \dfrac{8}{(2m + 1)^3 \pi^2}$.

EXERCISE (xv)

Obtain the solution $f(x, t)$ of the one-dimensional wave equation (14) which is of the form $f(x, t) = X(x) T(t)$ and for which $f(0, t) = e^{-3ct}$ and $f_x(2, t) = 0$ for all t. Show that the solution may be expressed in the form

$$f(x, t) = e^{-3ct} \frac{\cosh 3(2 - x)}{\cosh 6}.$$

EXERCISE (xvi)

Find a set of solutions of Laplace's equation of the form $f(x, y) = X(x) Y(y)$ for which $X(0) = 0 = X(1)$ and $Y(1) = 0$. Deduce that

$$f(x, y) = \sum_{n=1}^{\infty} C_n \sin n\pi x \sinh n\pi (1 - y)$$

is a solution which satisfies these boundary conditions. Find C_n if $f(x, 0) = \sin^2 \pi x$.

EXERCISE (xvii)

Use the method of separation of variables to obtain a solution of

$$\frac{\partial^2 V}{\partial x^2} - \frac{\partial V}{\partial y} = V$$

for which $V(x, y) = V(x + 2\pi, y)$, $V(0, y) = 0 = V(\pi, y)$ and $V(x, 1) = x$.

EXERCISE (xviii)

Express Laplace's equation in polar coordinates and find a solution of the form $\Sigma \varphi(r) \psi(\theta)$ which is bounded inside and takes the values $\cos 3\theta$ on the unit circle.

CHAPTER 7

Fourier Series

7.1 Definition of Fourier series

We saw in Chapter 4 how certain functions may be represented as the sum of a convergent infinite series. There the representation was given by a power series, the Taylor series expansion of the function, and conditions were given for convergence of the power series to the function in question.

Where the function $f(x)$ is *periodic*, say $f(x + 2l) = f(x)$ for all x (the *period* is then $2l$), there is another way of representing the function, namely by a *trigonometric series*. Such a series is one of the form

$$\tfrac{1}{2}a_0 + \sum_{n=1}^{\infty} \left(a_n \cos \frac{n\pi x}{l} + b_n \sin \frac{n\pi x}{l} \right) \tag{1}$$

or, when the period is 2π,

$$\tfrac{1}{2}a_0 + \sum_{n=1}^{\infty} (a_n \cos nx + b_n \sin nx). \tag{2}$$

Suppose then that $f(x)$ is defined in the interval $-l < x \leqslant l$, and that $f(x)$ is defined outside this interval by the assumption that it is periodic, of period $2l$. Suppose also that the following integrals exist

$$a_n = \frac{1}{l} \int_{-l}^{l} f(x) \cos \frac{n\pi x}{l}\, dx, \qquad n = 0, 1, 2, \ldots, \tag{3}$$

$$b_n = \frac{1}{l} \int_{-l}^{l} f(x) \sin \frac{n\pi x}{l}\, dx, \qquad n = 1, 2, 3, \ldots.. \tag{4}$$

(Note that these integrals may exist even if $f(x)$ is not continuous.) The numbers $a_0, a_1, a_2, \ldots, b_1, b_2, \ldots$ so defined are called the *Fourier coefficients* of the function $f(x)$. The series (1), in which the coefficients a_n and b_n are the Fourier coefficients of f, is then called the *Fourier series* of f.

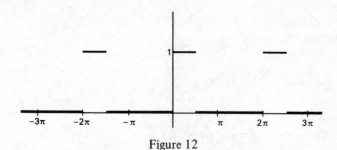

Figure 12

Example 1   Find the Fourier series of the function $f(x)$ (see Fig. 12) of period 2π where

$$f(x) = \begin{cases} 0 & \text{for } -\pi < x \leq 0, \\ 1 & \text{for } 0 < x \leq \pi/2, \\ 0 & \text{for } \pi/2 < x \leq \pi. \end{cases}$$

Solution   Using the definitions (3) and (4), we have

$$a_0 = \frac{1}{\pi} \int_{-\pi}^{\pi} f(x)\, dx = \frac{1}{\pi} \int_{0}^{\pi/2} dx = \tfrac{1}{2},$$

$$a_n = \frac{1}{\pi} \int_{-\pi}^{\pi} f(x) \cos nx\, dx = \frac{1}{\pi} \int_{0}^{\pi/2} \cos nx\, dx = \frac{1}{n\pi} \sin \frac{n\pi}{2},$$

for $n \neq 0$,

$$b_n = \frac{1}{\pi} \int_{-\pi}^{\pi} f(x) \sin nx\, dx = \frac{1}{\pi} \int_{0}^{\pi/2} \sin nx\, dx = \frac{1}{n\pi} \left(1 - \cos \frac{n\pi}{2} \right).$$

It follows then that $a_n = 0$ if n is even, $n \neq 0$, and

$$a_{2m+1} = \frac{(-1)^m}{(2m+1)\pi}.$$

Also

$$b_{2m} = \frac{1 - (-1)^m}{2m\pi},$$

$$b_{2m+1} = \frac{1}{(2m+1)\pi}.$$

Therefore the Fourier series of $f(x)$ is

$$\tfrac{1}{4} + \frac{1}{\pi}(\cos x + \sin x + \sin 2x - \tfrac{1}{3}\cos 3x + \tfrac{1}{3}\sin 3x + \tfrac{1}{5}\cos 5x$$

$$+ \tfrac{1}{5}\sin 5x + \tfrac{1}{3}\sin 6x - \tfrac{1}{7}\cos 7x + \tfrac{1}{7}\sin 7x + \cdots).$$

EXERCISE (i)
Show that

$$\sum_{n=1}^{\infty}(-1)^{n+1}\frac{2}{n}\sin nx$$

is the Fourier series of the function $f(x)$ of period 2π, where $f(x) = x$ for $-\pi < x \leqslant \pi$. Sketch the graph of this function in the range $-3\pi < x \leqslant 3\pi$.

EXERCISE (ii)
Show that

$$1 - \frac{8}{\pi^2}\sum_{n=0}^{\infty}\frac{1}{(2n+1)^2}\cos\frac{(2n+1)\pi x}{2}$$

is the Fourier series of $f(x)$ which has period 4 and where $f(x) = |x|$ for $-2 < x \leqslant 2$. Sketch the graph of $f(x)$ for $-6 < x \leqslant 6$.

Suppose that $f(x)$ and $g(x)$ take the same values throughout $-l < x \leqslant l$ except at a finite number of points. It follows from the definitions (3), (4) that the Fourier coefficients of f and g coincide, and therefore they have identical Fourier series.

So far we make no claim about the convergence of the Fourier series of $f(x)$. Even if it does converge there is no reason to suppose that its sum is in any way related to $f(x)$. For simplicity of notation we shall limit our consideration to functions $f(x)$ whose period is 2π, and whose Fourier series is then of the form (2). This is not a great restriction, for suppose that $g(x)$ is defined over $-l < x \leqslant l$. Define $f(x) = g(xl/\pi)$. Then $f(x)$ is defined over $-\pi < x \leqslant \pi$, and if g has period $2l$ then this implies that f has period 2π. Furthermore,

$$\frac{1}{\pi}\int_{-\pi}^{\pi}f(x)\cos nx\,dx = \frac{1}{l}\int_{-l}^{l}g(x)\cos\frac{n\pi x}{l}\,dx,$$

$$\frac{1}{\pi}\int_{-\pi}^{\pi}f(x)\sin nx\,dx = \frac{1}{l}\int_{-l}^{l}g(x)\sin\frac{n\pi x}{l}\,dx,$$

and so the question as to whether the series (1) converges to $g(x)$ is the same as asking whether (2) converges to $f(x)$.

7.2 Convergence

If we consider the function $f(x)$ which is defined to be -1 for all $x < 2$, and is equal to 1 whenever $x \geqslant 2$, then we readily conclude that $f(x)$ is continuous everywhere except at $x = 2$. At the point $x = 2$ the situation merits some investigation. The function $f(2 + h)$ does not tend to a limit as $h \to 0$. However, if we restrict the values of h to be *negative* and then allow h to tend to zero (i.e. let $2 + h$ tend to 2 from below), we see that $f(2 + h)$ tends to a limit, which is -1. Similarly, if h is restricted to be *positive* and $2 + h$ is allowed to tend to 2 from above, then $f(2 + h)$ tends to a limit, which is 1. More formally, we say

if $f(x)$ is defined for $\alpha < x < \beta$ we write
$$\lim_{x \to \beta - 0} f(x) = l \quad or \quad f(\beta - 0) = l$$

if l is a real number such that, given any $\epsilon < 0$, there exists $\delta > 0$ such that $|f(x) - l| < \epsilon$ whenever $0 < \beta - x < \delta$;

if $f(x)$ is defined for $\alpha < x < \beta$ we write
$$\lim_{x \to \alpha + 0} f(x) = l \quad or \quad f(\alpha + 0) = l$$

if l is a real number such that, given $\epsilon > 0$, there exists $\delta > 0$ such that $|f(x) - l| < \epsilon$ whenever $0 < x - \alpha < \delta$.

The values $f(\beta - 0), f(\alpha + 0)$ are called the *one-sided limits* of f at the points β and α. It follows from the definitions that f is *continuous* at x if $f(x - 0) = f(x) = f(x + 0)$. In our example above $f(2 - 0) = -1$, $f(2) = f(2 + 0) = 1$.

The function $f(x)$ is said to be *piecewise continuous* in the interval $\alpha \leqslant x \leqslant \beta$ if there is a finite subdivision of the interval

$$\alpha = x_0 < x_1 < \cdots < x_r = \beta \tag{5}$$

such that $f(x)$ is continuous on each subdivision $x_i < x < x_{i+1}$. We require further that the one-sided limits $f(x + 0), f(x - 0)$ exist at $x_1, \ldots, x_{r-1}$ and that $f(\alpha + 0)$ and $f(\beta - 0)$ exist. We say that the function *jumps* at each subdivision point x_i. Naturally, continuous functions are piecewise continuous, and the functions of Example 1 and Exs (i) and (ii) are piecewise continuous over any finite interval.

PROPOSITION 7.1 (cf. Proposition 3.7)
If $f(x)$ is piecewise continuous in $\alpha \leqslant x \leqslant \beta$ then it is bounded there.

Proof
Without loss of generality we may assume that $f(x)$ is continuous for $\alpha < x < \beta$, otherwise we prove the proposition for each subinterval

in which $f(x)$ is continuous. Now there exist δ_1, δ_2 such that $|f(\alpha) - f(x)| < 1$ whenever $0 < x - \alpha < \delta_1$, and $|f(\beta) - f(x)| < 1$ whenever $0 < \beta - x < \delta_2$. Hence $|f(x)| < 1 + |f(\alpha)|$ for $\alpha \leqslant x < \alpha + \delta_1$ and $|f(x)| < 1 + |f(\beta)|$ for $\beta - \delta_2 < x \leqslant \beta$. Since $|f(x)|$ is bounded for $\alpha + \delta_1 \leqslant x \leqslant \beta - \delta_2$ (Proposition 3.7) this proves the proposition.

The function $f(x)$ is said to be *piecewise smooth* in the interval $\alpha \leqslant x \leqslant \beta$ if both $f(x)$ and its first derivative $f'(x)$ are piecewise continuous in the interval.

We can now give sufficient conditions for a function to have convergent Fourier series.

PROPOSITION 7.2
Suppose that $f(x)$ is piecewise smooth on any finite interval and that $f(x)$ is periodic of period 2π. Then at each point x the Fourier series of f converges to

$$s(x) = \tfrac{1}{2}\{f(x + 0) + f(x - 0)\}.$$

In particular, at each point x at which the function is continuous, its Fourier series converges to $f(x)$.

The proof of this proposition depends on the following lemmas.

LEMMA 7.3
For every positive integer n and for all x such that $\sin \tfrac{1}{2}x \neq 0$ we have

$$\tfrac{1}{2} + \sum_{r=1}^{n} \cos rx = \frac{\sin (n + \tfrac{1}{2})x}{2 \sin \tfrac{1}{2}x},$$

$$\int_{0}^{\pi} \frac{\sin (n + \tfrac{1}{2})x}{2 \sin \tfrac{1}{2}x}\, dx = \frac{\pi}{2}.$$

Proof
Straightforward. Use $2 \sin \tfrac{1}{2}x \cos rx = \sin (r + \tfrac{1}{2})x - \sin (r - \tfrac{1}{2})x$, and sum both sides of the equality for $r = 1, 2, \ldots, n$.

LEMMA 7.4
If $s_n(x)$ denotes the partial sum

$$s_n(x) = \tfrac{1}{2}a_0 + \sum_{r=1}^{n} (a_r \cos rx + b_r \sin rx)$$

of the Fourier series of f, then

$$s_n(x) - \tfrac{1}{2}\{f(x+0) + f(x-0)\}$$

$$= \frac{1}{\pi} \int\limits_0^\pi \{f(x+t) - f(x+0)\}D_n(t)\,dt$$

$$+ \frac{1}{\pi} \int\limits_0^\pi \{f(x-t) - f(x-0)\}D_n(t)\,dt \tag{6}$$

where $D_n(t) = (\sin(n + \tfrac{1}{2})t)/(2 \sin \tfrac{1}{2}t)$.

Proof
From the definition of the Fourier coefficients we have

$$s_n(x) = \frac{1}{\pi} \int\limits_{-\pi}^\pi f(y)\left\{\tfrac{1}{2} + \sum_{r=1}^n (\cos ry \cos rx + \sin ry \sin rx)\right\} dy$$

$$= \frac{1}{\pi} \int\limits_{-\pi}^\pi f(y)\left\{\tfrac{1}{2} + \sum_{r=1}^n \cos r(y-x)\right\} dy.$$

We change the variable of integration by setting $y = x + t$, and obtain

$$s_n(x) = \frac{1}{\pi} \int\limits_{-\pi-x}^{\pi-x} f(x+t)\left\{\tfrac{1}{2} + \sum_{r=1}^n \cos rt\right\} dt$$

$$= \frac{1}{\pi} \int\limits_{-\pi}^{\pi} f(x+t)\left\{\tfrac{1}{2} + \sum_{r=1}^n \cos rt\right\} dt$$

since both f and $\tfrac{1}{2} + \Sigma_{r=1}^n \cos rt$ are periodic of period 2π. [We use here the theorem that, if $g(t)$ is periodic of period 2π then $\int_{-\pi-x}^{\pi-x} g(t)\,dt$ is independent of x. Prove this.] Therefore, using Lemma 7.3 and noting that $D_n(t) = D_n(-t)$, we see that

$$s_n(x) = \frac{1}{\pi} \int\limits_0^\pi f(x+t)\,D_n(t)\,dt + \frac{1}{\pi} \int\limits_0^\pi f(x-t)\,D_n(t)\,dt$$

and

$$\tfrac{1}{2}\{f(x+0) + f(x-0)\} = \frac{1}{\pi} \int\limits_0^\pi f(x+0) D_n(t)\, dt$$

$$+ \frac{1}{\pi} \int\limits_0^\pi f(x-0) D_n(t)\, dt.$$

Together these yield (6).

LEMMA 7.5
If the function $g(x)$ is piecewise continuous in the interval $\alpha \leqslant x \leqslant \beta$, then the integral

$$I(\lambda) = \int\limits_\alpha^\beta g(t) \sin \lambda t\, dt$$

tends to zero as $\lambda \to \infty$.

Proof
For the purposes of the proof we may as well assume that $g(x)$ is continuous in $\alpha < x < \beta$, since otherwise we need only prove the lemma for each subinterval in which $g(x)$ is continuous. Let M denote an upper bound for $|g(x)|$ in $\alpha \leqslant x \leqslant \beta$. If we change the variable of integration by taking $u = t - \rho$, where $\rho = \pi/\lambda$, then we obtain

$$I(\lambda) = \int\limits_{\alpha-\rho}^{\beta-\rho} g(u+\rho) \sin \lambda(u+\rho)\, du$$

$$= \int\limits_{\alpha-\rho}^{\beta-\rho} g(u+\rho) \sin (\lambda u + \pi)\, du = - \int\limits_{\alpha-\rho}^{\beta-\rho} g(u+\rho) \sin \lambda u\, du.$$

Therefore

$$2I(\lambda) = \int\limits_\alpha^\beta g(t) \sin \lambda t\, dt - \int\limits_{\alpha-\rho}^{\beta-\rho} g(t+\rho) \sin \lambda t\, dt$$

$$= \int\limits_{\alpha-\rho}^\alpha g(t+\rho) \sin \lambda t\, dt + \int\limits_\alpha^{\beta-\rho} \{g(t) - g(t+\rho)\} \sin \lambda t\, dt$$

$$+ \int\limits_{\beta-\rho}^\beta g(t) \sin \lambda t\, dt.$$

We have then

$$2\,|\,I(\lambda)\,| \leqslant 2M\rho + \int\limits_{\alpha}^{\beta-\rho} |\,g(t) - g(t+\rho)\,|\,dt.$$

Given $\epsilon > 0$, we can choose λ so large (i.e. ρ so small) that throughout $\alpha \leqslant t \leqslant \beta - \rho$

$$|\,g(t) - g(t+\rho)\,| < \epsilon/(\beta - \alpha)*$$

and also

$$2M\rho < \epsilon,$$

which implies that $|\,I(\lambda)\,| < \epsilon$. Therefore $I(\lambda) \to 0$ as $\lambda \to \infty$.

(* This is because, in a closed interval, a continuous function is *uniformly continuous*. The proof of this assertion is beyond the scope of this book.)

Proof of Proposition 7.2

It remains to show that, under the hypotheses of the proposition, the integrals on the right-hand side of (6) tend to zero as $n \to \infty$. This follows directly from Lemma 7.5, provided we can show that the functions

$$\frac{f(x+t) - f(x+0)}{2 \sin \frac{1}{2}t}, \qquad \frac{f(x-t) - f(x-0)}{2 \sin \frac{1}{2}t} \qquad (7)$$

are piecewise continuous in $0 \leqslant t \leqslant \pi$. Both functions are clearly piecewise continuous for $0 < t \leqslant \pi$, and the only problem occurs near $t = 0$. Since $f'(x+0), f'(x-0)$ exist then

$$\lim_{t \to 0+} \frac{f(x+t) - f(x+0)}{2 \sin \frac{1}{2}t} = \lim_{t \to 0+} \frac{f(x+t) - f(x+0)}{t} \frac{\frac{1}{2}t}{\sin \frac{1}{2}t}$$

$$= f'(x+0),$$

$$\lim_{t \to 0+} \frac{f(x-t) - f(x-0)}{2 \sin \frac{1}{2}t} = \lim_{t \to 0+} \frac{f(x-t) - f(x-0)}{t} \frac{\frac{1}{2}t}{\sin \frac{1}{2}t}$$

$$= f'(x-0).$$

This verifies that both functions (7) tend to a limit as $t \to 0+$, and we therefore have piecewise continuity throughout $0 \leqslant t \leqslant \pi$.

7.3 Applications

Example 2 Use the result of Ex. (i) to show that

$$\frac{\pi}{4} = 1 - \frac{1}{3} + \frac{1}{5} - \cdots + \frac{(-1)^{n+1}}{2n-1} + \cdots.$$

Solution   From the graph of this function we see that the only discontinuities occur at odd multiples of π. Hence the function is continuous at $x = \pi/2$ and its Fourier series $\Sigma(-1)^{n+1}(2/n)\sin nx$ converges to $\pi/2$ there.

Therefore

$$\frac{\pi}{4} = \sum_{n=1}^{\infty} (-1)^{n+1} \frac{1}{n} \sin \frac{n\pi}{2} = \sum_{n=1}^{\infty} \frac{(-1)^{n+1}}{2n-1}.$$

EXERCISE (iii)

Find the Fourier series of the function $f(x)$ of period 4, where $f(x) = x^2$ for $-2 < x \leqslant 2$. By considering the convergence of the Fourier series at $x = 0$, or otherwise, show that

$$\frac{\pi^2}{12} = 1 - \frac{1}{2^2} + \frac{1}{3^2} - \frac{1}{4^2} + \cdots + (-1)^{n+1} \frac{1}{n^2} + \cdots.$$

EXERCISE (iv)

Use the results of Exs (ii) and (iii) to show that

$$\frac{\pi^2}{8} = 1 + \frac{1}{3^2} + \frac{1}{5^2} + \cdots + \frac{1}{(2n-1)^2} + \cdots,$$

and

$$\frac{\pi^2}{6} = 1 + \frac{1}{2^2} + \frac{1}{3^2} + \frac{1}{4^2} + \cdots + \frac{1}{n^2} + \cdots.$$

Example 3   Find the Fourier series of $f(x) = e^{\lambda x} (\lambda \neq 0)$ for $-\pi < x \leqslant x$, where $f(x)$ has period 2π. Deduce from this the Fourier series of $\cosh \lambda x$ and $\sinh \lambda x$, and show that

$$\sum_{n=1}^{\infty} \frac{1}{\lambda^2 + n^2} = \frac{1}{2\lambda^2} (\pi\lambda \coth \lambda\pi - 1).$$

Solution

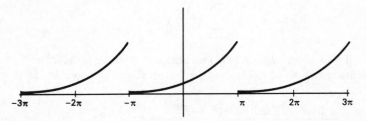

Figure 13

We have

$$a_0 = \frac{1}{\pi} \int\limits_{-\pi}^{\pi} e^{\lambda x} \, dx = \frac{2}{\lambda\pi} \sinh \lambda\pi,$$

$$a_n + ib_n = \frac{1}{\pi} \int\limits_{-\pi}^{\pi} e^{\lambda x} (\cos nx + i \sin nx) \, dx = \frac{1}{\pi} \int\limits_{-\pi}^{\pi} e^{(\lambda + in)x} \, dx$$

$$= \frac{2}{\pi(\lambda + in)} \sinh (\lambda + in)\pi = \frac{2}{\pi(\lambda + in)} (-1)^n \sinh \lambda\pi,$$

since

$$\sinh (\lambda + in)\pi = \sinh \lambda\pi \cosh in\pi + \cosh \lambda\pi \sinh in\pi$$
$$= \sinh \lambda\pi \cos n\pi + i \cosh \lambda\pi \sin n\pi = (-1)^n \sinh \lambda\pi.$$

Therefore

$$a_n + ib_n = \frac{2(-1)^n \sinh \lambda\pi}{\pi(\lambda^2 + n^2)} (\lambda - in),$$

and so the Fourier series is

$$\frac{\sinh \lambda\pi}{\lambda\pi} + \frac{2 \sinh \lambda\pi}{\pi} \sum_{n=1}^{\infty} \frac{(-1)^n}{\lambda^2 + n^2} \{\lambda \cos nx - n \sin nx\}. \tag{8}$$

For $-\pi < x < \pi$, the series (8) converges to $e^{\lambda x}$, and for $x = \pi$ (8) converges to $\frac{1}{2}\{e^{\lambda\pi} + e^{-\lambda\pi}\}$, i.e. to $\cosh \lambda\pi$. Therefore

$$\cosh \lambda\pi = \frac{\sinh \lambda\pi}{\lambda\pi} + \frac{2 \sinh \lambda\pi}{\pi} \sum_{n=1}^{\infty} \frac{(-1)^n}{\lambda^2 + n^2} \lambda \cos n\pi,$$

and so

$$\lambda\pi \coth \lambda\pi - 1 = 2\lambda^2 \sum_{n=1}^{\infty} \frac{1}{\lambda^2 + n^2} \, .$$

If we replace λ *by* $-\lambda$ in (8) we obtain the Fourier series for $e^{-\lambda x}$ which, when added to (8) gives the Fourier series for $2 \cosh \lambda x$. Thus

$$\frac{\sinh \lambda\pi}{\lambda\pi} + \frac{2\lambda \sinh \lambda\pi}{\pi} \sum_{n=1}^{\infty} \frac{(-1)^n}{\lambda^2 + n^2} \cos nx \tag{9}$$

and (by subtracting from (8))

$$\frac{2 \sinh \lambda \pi}{\pi} \sum_{n=1}^{\infty} \frac{(-1)^{n+1} n}{\lambda^2 + n^2} \sin nx \qquad (10)$$

are the Fourier series of $\cosh \lambda x$ and $\sinh \lambda x$ respectively.

EXERCISE (v)

If $f(x) = f(-x)$ (i.e. f is an *even* function) and f is periodic of period $2l$, show that the Fourier coefficients $b_n = 0$ for $n = 1, 2, 3, \ldots$ and that f can be expanded in a *cosine series*. Verify that $f(x) = |\sin x|$ is an even function of period π and deduce that

$$|\sin x| = \frac{2}{\pi} \left(1 - 2 \sum_{n=1}^{\infty} \frac{\cos 2nx}{4n^2 - 1} \right).$$

EXERCISE (vi)

If $f(-x) = -f(x)$ (i.e. f is an *odd* function) and f is periodic of period $2l$, show that the Fourier coefficients $a_n = 0$ for $n = 0, 1, 2, \ldots$ and that f can be expanded in a *sine series*. Sketch the graph of the function $f(x)$ of period 2π, where $f(x) = \frac{1}{2}(\pi - x)$ for $0 < x \leq 2\pi$ and verify that f is an odd function and that $\frac{1}{2}(\pi - x) = \Sigma n^{-1} \sin nx$.

EXERCISE (vii)

A function $f(x)$ is defined in the range $-5 \leq x \leq 5$ by

$$f(x) = \begin{cases} 0 & \text{for } -5 < x < 0 \\ 2 & \text{for } 0 < x < 5 \end{cases}$$

and $f(-5) = f(0) = f(5) = 1$. Show that its Fourier series over the range in question is

$$1 + \frac{4}{\pi} \left(\sin \frac{\pi x}{5} + \frac{1}{3} \sin \frac{3\pi x}{5} + \frac{1}{5} \sin \frac{5\pi x}{5} + \cdots \right).$$

Derive from this a series $F(x, t)$ of the form

$$\sum_{n=1}^{\infty} V_n(t) \sin \frac{n\pi x}{5}$$

which satisfies, for $t \geq 0$, the equation

$$\frac{\partial^2 F}{\partial x^2} = \frac{\partial F}{\partial t}$$

and the boundary conditions

$$F(0, t) = F(5, t) = 0 \quad \text{for all } t \geqslant 0,$$
$$F(x, 0) = f(x) - 1 \quad \text{for } -5 \leqslant x \leqslant 5.$$

7.4 Integration and differentiation of Fourier series

PROPOSITION 7.6

Suppose that $f(x)$ is piecewise continuous in $-\pi \leqslant x \leqslant \pi$ and that its Fourier series is given by (2). Then for $-\pi \leqslant \alpha < \beta \leqslant \pi$ we have

$$\int_{\alpha}^{\beta} f(x)\, dx = \tfrac{1}{2} a_0 (\beta - \alpha) + \sum_{n=1}^{\infty} \int_{\alpha}^{\beta} (a_n \cos nx + b_n \sin nx)\, dx.$$

(Note: we do not assume that the Fourier series of f converges to $f(x)$, in fact we do not assume that it converges at all!)

Proof

Let $F(x)$ be defined by the equation

$$F(x) = \int_{-\pi}^{x} \{f(t) - \tfrac{1}{2} a_0\}\, dt, \qquad -\pi < x \leqslant \pi,$$

F being of period 2π. Since $F(\pi) = 0$, then $F(x)$ is continuous at π, and hence everywhere. Moreover it is piecewise smooth, and it follows then that the Fourier series of $F(x)$

$$\tfrac{1}{2} A_0 + \sum_{n=1}^{\infty} (A_n \cos nx + B_n \sin nx)$$

converges to $F(x)$. Now for $n \neq 0$,

$$A_n + i B_n = \frac{1}{\pi} \int_{-\pi}^{\pi} F(x)\, e^{inx}\, dx$$

$$= \frac{1}{in\pi} [F(x)\, e^{inx}]_{-\pi}^{\pi} - \frac{1}{in\pi} \int_{-\pi}^{\pi} e^{inx} \{f(x) - \tfrac{1}{2} a_0\}\, dx$$

$$= \frac{i}{n\pi} \int_{-\pi}^{\pi} e^{inx} f(x)\, dx = \frac{i}{n} (a_n + i b_n).$$

It follows then that $A_n = -b_n/n$, $B_n = a_n/n$, and so

$$F(\beta) - F(\alpha) = \sum_{n=1}^{\infty} \left\{ \frac{a_n}{n} (\sin n\beta - \sin n\alpha) - \frac{b_n}{n} (\cos n\beta - \cos n\alpha) \right\}.$$

This implies that

$$\int_{\alpha}^{\beta} f(x)\, dx - \tfrac{1}{2} a_0(\beta - \alpha) = F(\beta) - F(\alpha)$$

$$= \sum_{n=1}^{\infty} \left\{ a_n \int_{\alpha}^{\beta} \cos nx\, dx + b_n \int_{\alpha}^{\beta} \sin nx\, dx \right\}.$$

Example 4   Find the Fourier series of $f(x) = |x|$, $-\pi < x \leqslant \pi$ and by integrating over a suitable interval find the sum of

$$1 - \frac{1}{3^3} + \frac{1}{5^3} - \frac{1}{7^3} + \cdots.$$

Solution $f(x)$ is an even function, and so $b_n = 0$ for all n. Now

$$a_0 = \frac{2}{\pi} \int_0^{\pi} x\, dx = \pi,$$

$$a_n = \frac{2}{\pi} \int_0^{\pi} x \cos nx\, dx = \frac{-2}{n^2\pi} (1 - (-1)^n).$$

Therefore for $-\pi < x \leqslant \pi$,

$$|x| = \frac{\pi}{2} - \frac{4}{\pi} \sum_{n=0}^{\infty} \frac{1}{(2n+1)^2} \cos(2n+1)x.$$

Integrating over 0 to $\pi/2$, we obtain

$$\frac{\pi^2}{8} = \frac{\pi^2}{4} - \frac{4}{\pi} \sum_{n=0}^{\infty} \frac{1}{(2n+1)^3} \sin \frac{(2n+1)\pi}{2}$$

i.e.

$$\frac{\pi^3}{32} = 1 - \frac{1}{3^3} + \frac{1}{5^3} - \frac{1}{7^3} + \cdots.$$

EXERCISE (viii)

Use the result of Example 4 to show that the Fourier series of $g(x)$ of period 2π, and where

$$g(x) = \begin{cases} \dfrac{x^2 - \pi x}{2} & \text{for } 0 \leqslant x \leqslant \pi, \\[2mm] \dfrac{-x^2 - \pi x}{2} & \text{for } -\pi \leqslant x < 0. \end{cases}$$

is

$$-\frac{4}{\pi} \sum_{n=0}^{\infty} \frac{\sin (2n + 1)x}{(2n + 1)^3} .$$

Hence show that the sum of the series $1 + 1/3^4 + 1/5^4 + 1/7^4 + \cdots$ is $\pi^4/96$, and deduce that $1 + 1/2^4 + 1/3^4 + 1/4^4 + \cdots$ has sum $\pi^4/90$.

EXERCISE (ix)

Find the Fourier series of the even function $f(x)$ which has period 2π and for which $f(x) = 2 \cos x$ when $0 < x \leqslant \pi/2$ and $f(x) = 0$ when $\pi/2 < x \leqslant \pi$. By integrating this series over a suitable range, show that the series

$$\frac{1}{1.2.3} - \frac{1}{5.6.7} + \frac{1}{9.10.11} - \cdots$$

has sum $\pi(\sqrt{2} - 1)/8$.

PROPOSITION 7.7

Suppose that $f(x)$ is continuous for all x and is periodic with period 2π. Suppose also that $f'(x)$ is piecewise continuous in any finite interval. Then the Fourier series of f' is obtained by differentiating that of f term by term.

Proof

Let the Fourier coefficients for $f(x)$ be denoted by $a_0, a_1, a_2, \ldots$ and $b_1, b_2, b_3, \ldots$, while those for $f'(x)$ are denoted by $A_0, A_1, A_2, \ldots$ and $B_1, B_2, B_3, \ldots$. Then we are required to show that $A_0 = 0$ and $A_n = nb_n, B_n = -na_n$. We have, for $n \neq 0$,

$$a_n + ib_n = \frac{1}{\pi} \int\limits_{-\pi}^{\pi} f(x)\, e^{inx}\, dx$$

$$= \frac{1}{in\pi} \{ e^{in\pi} f(\pi) - e^{-in\pi} f(-\pi) \} - \frac{1}{in\pi} \int\limits_{-\pi}^{\pi} e^{inx} f'(x)\, dx$$

$$= \frac{i}{n\pi} \int\limits_{-\pi}^{\pi} e^{inx} f'(x)\, dx \quad \text{(since } f(\pi) = f(-\pi) \text{)}$$

$$= \frac{i}{n} (A_n + iB_n).$$

Finally,

$$A_0 = \frac{1}{\pi} \{ f(\pi) - f(-\pi) \} = 0.$$

EXERCISE (x)
Use the result of Example 4 to find the Fourier series of the odd function $f(x)$ of period 2π where $f(x) = 1$ for $0 < x < \pi$.

EXERCISE (xi)
By using the result of Ex. (v), find a trigonometric series whose sum is $\cos x$ when $0 < x < \pi$, and $-\cos x$ when $-\pi < x < 0$.

Index